Everyday Mathematics®

Student Math Journal 2

The University of Chicago
School Mathematics Project

The McGraw-Hill Companies

UCSMP Elementary Materials Component

Max Bell, Director

Authors

Max Bell
Jean Bell
John Bretzlauf*
Amy Dillard*
Robert Hartfield
Andy Isaacs*
James McBride, Director
Kathleen Pitvorec*
Peter Saecker

Technical Art

Diana Barrie*

**Second Edition only*

Contributors

Carol Arkin, Robert Balfanz, Sharlean Brooks, Ellen Dairyko, James Flanders, David Garcia, Rita Gronbach, Deborah Arron Leslie, Curtis Lieneck, Diana Marino, Mary Moley, William D. Pattison, William Salvato, Jean Marie Sweigart, Leeann Wille

Photo Credits

Phil Martin/Photography, Jack Demuth/Photography, Cover Credits: Bee/Stephen Dalton/Photo Researchers Inc., Photo Collage: Herman Adler Design Group

Send all inquiries to:
Wright Group/McGraw-Hill
P.O. Box 812960
Chicago, IL 60681

Printed in the United States of America.

ISBN 0-07-584463-X

16 17 DBH 10 09 08 07

Contents

Unit 7: Patterns and Rules

A note at the bottom of each journal page indicates when that page is first used. Some pages will be used again during the course of the year.

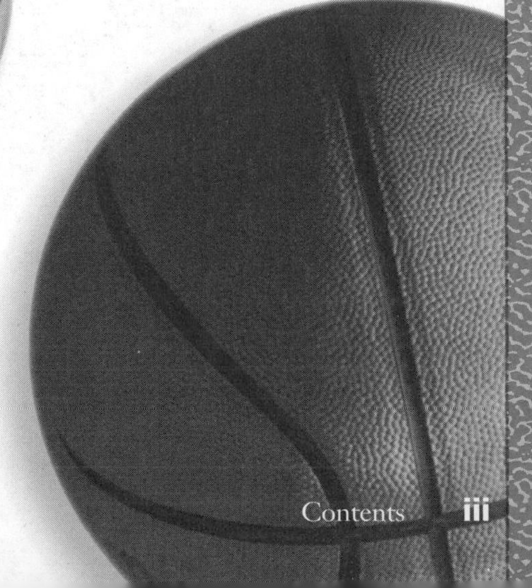

Unit 8: Fractions

Unit 9: Measurement

Unit 10: Decimals and Place Value

Unit 11: Whole Number Operations Revisited

Unit 12: Year-End Reviews and Extensions

Activity Sheets

Number Patterns

1. Count by 5s starting with the number 102. Color in the numbers on the grid with a crayon. Can you find a pattern?

									100
101	102	103	104	105	106	107	108	109	110
111	112	113	114	115	116	117	118	119	120
121	122	123	124	125	126	127	128	129	130

2. Pick a number to start with. Pick a number to count by. Mark your counts using a crayon.

									300
301	302	303	304	305	306	307	308	309	310
311	312	313	314	315	316	317	318	319	320
321	322	323	324	325	326	327	328	329	330
331	332	333	334	335	336	337	338	339	340
341	342	343	344	345	346	347	348	349	350
351	352	353	354	355	356	357	358	359	360
361	362	363	364	365	366	367	368	369	370

I counted by ______ starting with the number ______.

I used a ________________ crayon to mark the counts.
(color)

Here is the pattern I found:

____

Date Time

Math Boxes 7.1

1. Write a number model for a ballpark estimate. Then subtract.

$$\begin{array}{r} 81 \\ -\ 22 \\ \hline \end{array}$$

Answer

Ballpark estimate:

____

2. Share 1 dozen cookies equally among 5 children. Draw a picture.

Each child gets ______ cookies.

There are ______ cookies left over.

3. 8 books per shelf. 4 shelves. Fill in the multiplication diagram and solve.

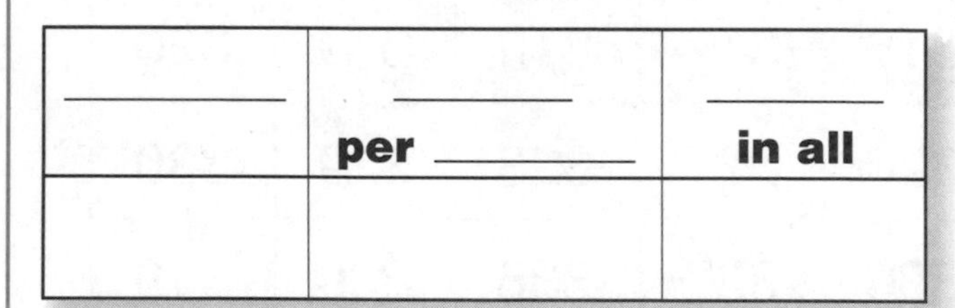

______	______ per ______	______ in all

There are ______ books.

4. Measure the line segment.

____

about ______ in.

about ______ cm

5. 15 dogs.
13 cats.
12 birds.

How many animals?

______ animals

6.

1 hour earlier is ______ : ______.

1 hour later is ______ : ______.

Date Time

Number Grids and Arrow Paths

Complete the pieces of the number grid. Solve the arrow-path puzzles.

1. 551

Start

551

End

2. 83

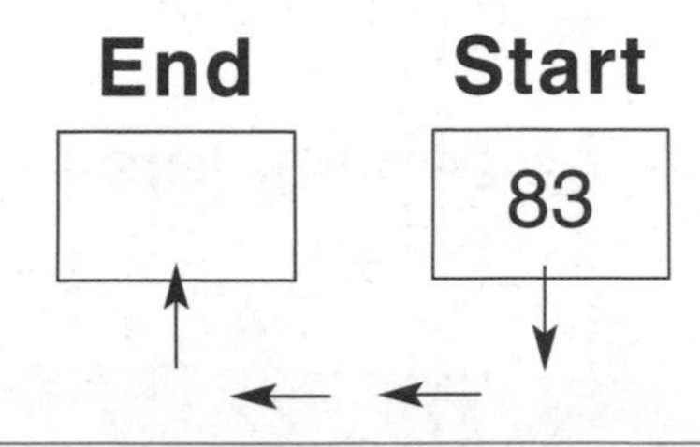

3. 797

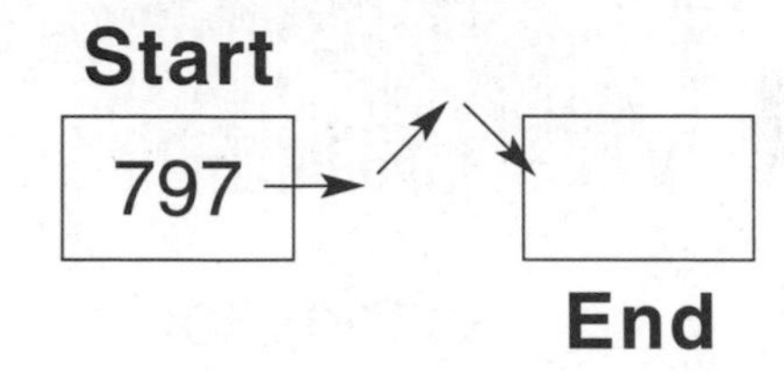

4.

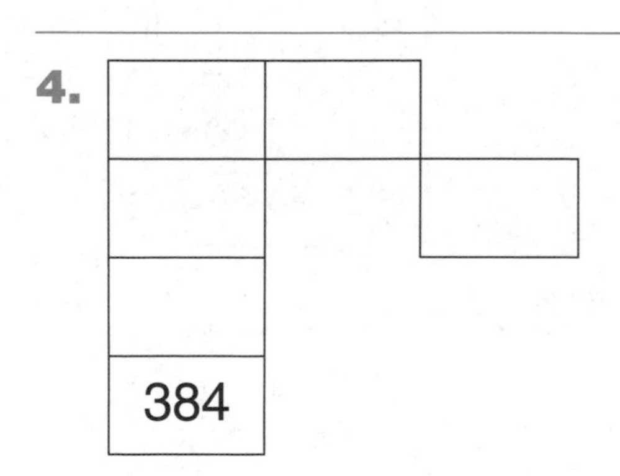

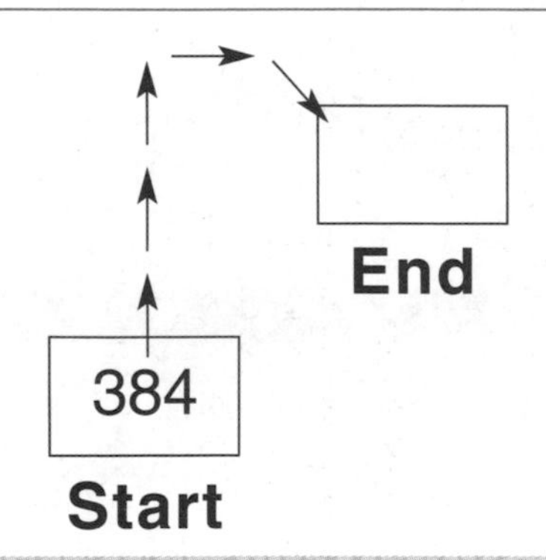

Challenge

5. Start

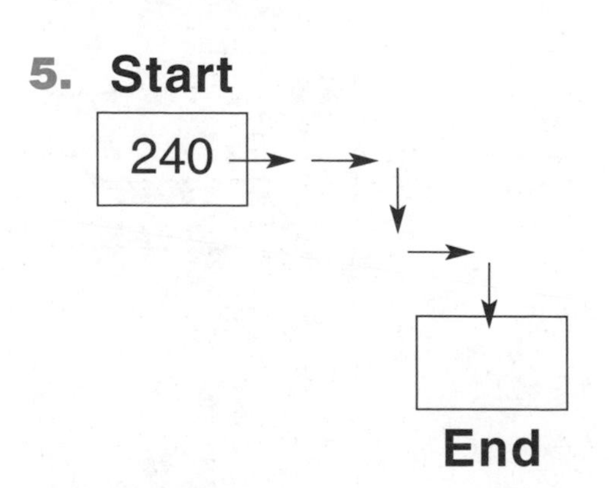

6.

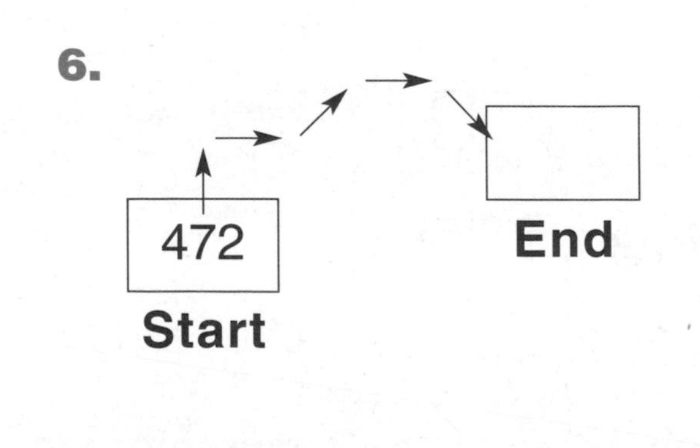

Date Time

Measuring Lengths with a Ruler

Use your ruler to measure the length of each object to the nearest inch and the nearest centimeter.

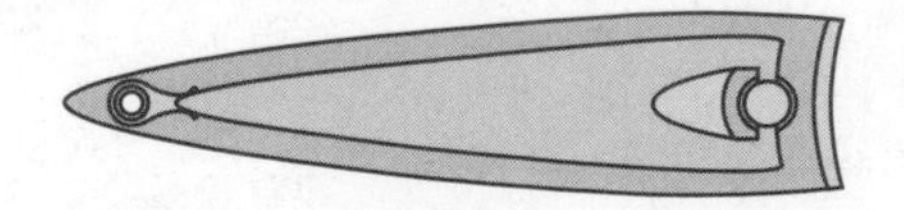

1. nail clipper

about ______ inches long

about ______ centimeters long

2. crayon

about ______ inches long

about ______ centimeters long

3. comb

about ______ inches long

about ______ centimeters long

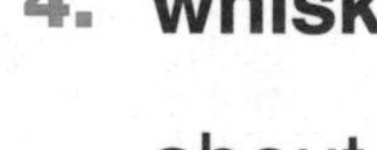

4. whisk

about ______ inches long

about ______ centimeters long

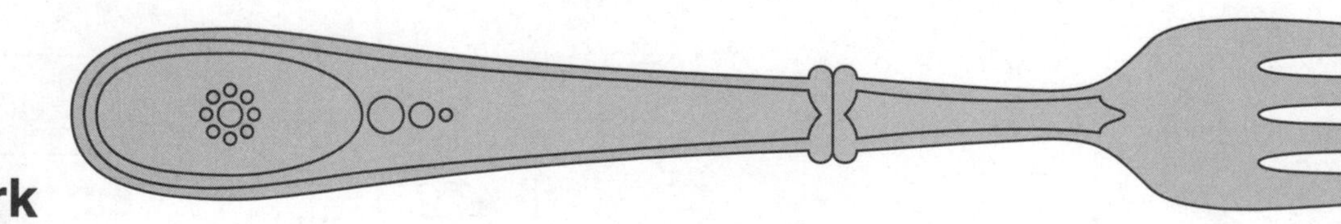

5. fork

about ________ inches long

about ________ centimeters long

Date Time

Math Boxes 7.2

1. Continue.

312, 314, 316, _____, _____,

_____, _____, _____, _____

2. Draw 5 nests with 3 eggs in each.

How many eggs in all?

_____ eggs

3. Solve.

Unit
children

70 − 20 = _____

_____ = 90 + 40

_____ = 120 − 80

150 − 60 = _____

4. Fill in the missing numbers.

	217	

5. Use your calculator. Start at 92. Count by 5s.

92, 97, _____, _____, _____,

_____, _____, _____, _____

6. Use the partial-sums algorithm to add. Show your work.

$$\begin{array}{r} 35 \\ +\ 79 \\ \hline \end{array}$$

Date Time

Making 10s

Record three rounds of *Hit the Target.*

Example Round

Target Number: 40

Starting Number	Change ↷	Result	Change ↷	Result	Change ↷	Result
12	+38	50	−10	40		

Round 1

Target Number: ______

Starting Number	Change ↷	Result	Change ↷	Result	Change ↷	Result

Round 2

Target Number: ______

Starting Number	Change ↷	Result	Change ↷	Result	Change ↷	Result

Round 3

Target Number: ______

Starting Number	Change ↷	Result	Change ↷	Result	Change ↷	Result

Date Time

Trade-First Subtraction Practice

For each problem, do the following:

- Before subtracting, write a number model for your ballpark estimate.
- If your estimate is 50 or less, subtract the numbers. Write your answer in the answer box.
- If your estimate is more than 50, you do not have to find an exact answer. Leave the answer box empty.

1. $74 - 31$ Answer: ____ Ballpark estimate: ____	**2.** $57 - 28$ Answer: ____ Ballpark estimate: ____	**3.** $84 - 17$ Answer: ____ Ballpark estimate: ____
4. $56 - 18$ Answer: ____ Ballpark estimate: ____	**5.** $87 - 25$ Answer: ____ Ballpark estimate: ____	**6.** $93 - 75$ Answer: ____ Ballpark estimate: ____
7. $84 - 27$ Answer: ____ Ballpark estimate: ____	**8.** $76 - 17$ Answer: ____ Ballpark estimate: ____	**9.** $53 - 29$ Answer: ____ Ballpark estimate: ____

Date Time

Math Boxes 7.3

1. Solve. Unit 62 + _____ = 70 50 = 44 + _____ 45 + _____ = 50 140 = 133 + _____	**2.** Write 5 names for 130. **130**
3. Use counters to solve. $14.00 is shared equally. Each child gets $5.00. How many children are sharing? _________ children How many dollars are left over? _________	**4.** Show 96¢ in two different ways. Use Ⓟ, Ⓝ, Ⓓ, and Ⓠ. _______________ _______________ _______________
5. Write the number that is 10 less. 65 _______ 400 _______ 260 _______ 1,391 _______	**6.** Room A has 35 desks. Room B has 29 desks. How many desks in all? _____ desks

Date Time

Basketball Addition

	Points Scored			
	Team 1		**Team 2**	
	1st Half	**2nd Half**	**1st Half**	**2nd Half**
Player 1				
Player 2				
Player 3				
Player 4				
Player 5				
Team Score				

Point Totals	**1st Half**	**2nd Half**	**Final**
Team 1	______	______	______
Team 2	______	______	______

1. Which team won the first half? ______________

By how much? ______ points

2. Which team won the second half? ______________

By how much? ______ points

3. Which team won the game? ______________

By how much? ______ points

Date Time

Math Boxes 7.4

1. 24 children. 6 in each row. Draw a picture.

How many rows? _____ rows

How many children left over?

_____ children

2. Solve the arrow-path puzzle.

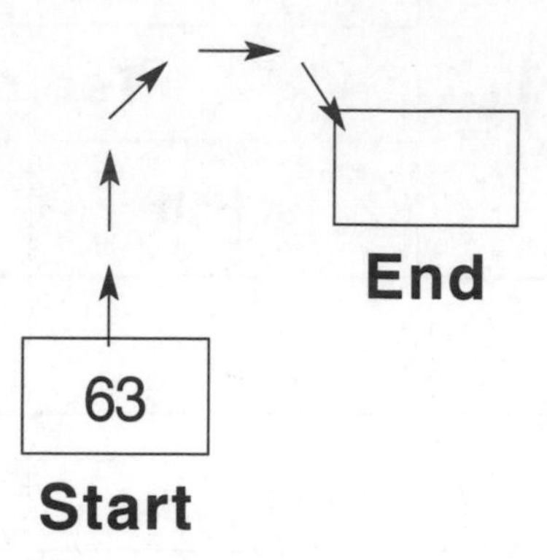

3. Write a number model for a ballpark estimate. Solve.

$$\begin{array}{r} 68 \\ +\ 35 \\ \hline \end{array}$$

Answer

Ballpark estimate:

____

4. Solve.

Unit
birds

$10 - \_\_\_\_\_ = 8$

$\_\_\_\_\_ - 6 = 14$

$33 = \_\_\_\_\_ - 7$

$90 - \_\_\_\_\_ = 82$

5. 18 cans of juice are shared by 5 people. Draw a picture.

_____ cans per person

_____ cans left over

6. Fill in the diagram and write a number model.

Start	Change	End
35	+35	

____

Date Time

The Wubbles

1. On each line, write the number of Wubbles after doubling. Use your calculator to help you.

You started on Friday with ______ Wubble.

On Saturday, there were ______ Wubbles.

On Sunday, there were ______ Wubbles.

On Monday, there will be ______ Wubbles.

On Tuesday, there will be ______ Wubbles.

On Wednesday, there will be ______ Wubbles.

On Thursday, there will be ______ Wubbles.

On Friday, there will be ______ Wubbles.

A Wubble

---

2. On each line, write the number of Wubbles after halving. Use your calculator to help you. Remember that "$\frac{1}{2}$ of " means "divide by 2."

There were ______ Wubbles.

After Wink 1, there were ______ Wubbles.

After Wink 2, there were ______ Wubbles.

After Wink 3, there were ______ Wubbles.

After Wink 4, there were ______ Wubbles.

After Wink 5, there were ______ Wubbles.

After Wink 6, there were ______ Wubbles.

After Wink 7, there was ______ Wubble.

Your room could look like this! What will you do?

Adapted with permission from *Calculator Mathematics Book 2* by Sheila Sconiers, pp. 10 and 11 (Everyday Learning Corporation, © 1990 by the University of Chicago).

Date Time

Math Boxes 7.5

1. The Jays scored 63 points. The Gulls scored 46 points. How many more points did the Jays score? ____ more points

Fill in the diagram.

Quantity	
Quantity	**Difference**

2. Solve the arrow-path puzzle.

Start 116 → ↓ ↘ → ↑ ↗ **End** ____

3. Solve.

Unit
train cars

$4 + 3 + 13 =$ ____

____ $= 12 + 6 + 8$

$5 + 4 + 18 =$ ____

____ $= 18 + 12 + 6$

$40 = 15 + 6 +$ ____

4. Write a number model for a ballpark estimate. Subtract.

$$\begin{array}{r} 86 \\ -\ 49 \\ \hline \end{array}$$

Answer

Ballpark estimate:

____

5. Solve.

Unit
pages read

____ $- 13 = 17$

$40 -$ ____ $= 28$

$39 =$ ____ $- 21$

$64 = 90 -$ ____

6. In Pensacola, Florida, the temperature is 82°F. In Portland, Maine, the temperature is 64°F. What is the difference?

____

Math Boxes 7.6

1. In basketball you can score 2 points for a basket. Josh made eight 2-point baskets. How many points did he score in all?

   _____ points

2. Solve.

   Unit
   missing cards

   17 − 9 = ____

   27 − 9 = ____

   57 − 9 = ____

   ____ = 77 − 9

   ____ = 97 − 9

3. Use counters to make a 5-by-2 array. Draw the array.

   How many counters in all?

   _____ counters

4. Write the number that is 100 less.

   465 ______

   700 ______

   960 ______

   4,391 ______

5. Double.

   Unit

   2 ______

   4 ______

   10 ______

   50 ______

6. Find the arrow rules.

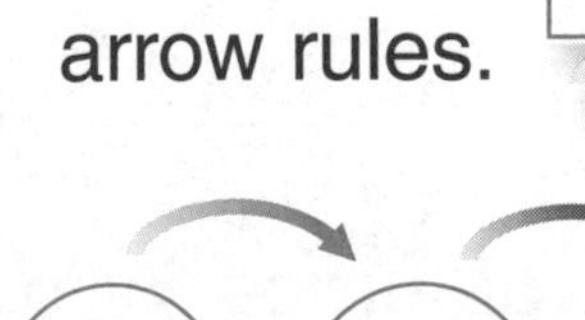

356

Date Time

Record of Our Jumps

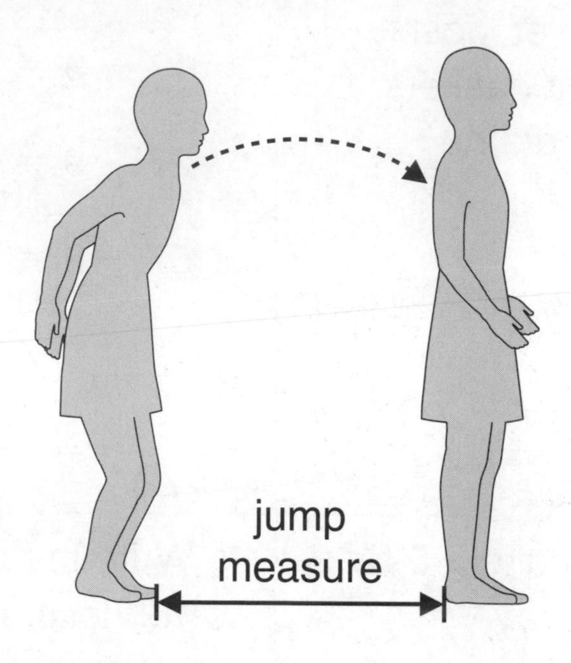

Place a penny or other marker, or make a dot with chalk, where the Jumper's back heel lands. Measure from the starting line to the marker. The jumps are measured to the nearest centimeter.

1. Record two of your jumps. Measure jumps to the nearest centimeter.

   First try: ______ centimeters

   Second try: ______ centimeters

2. My longer jump was ______ centimeters.

3. A middle value of jumps for our class is ______ centimeters.

Date Time

Record of Our Arm Spans

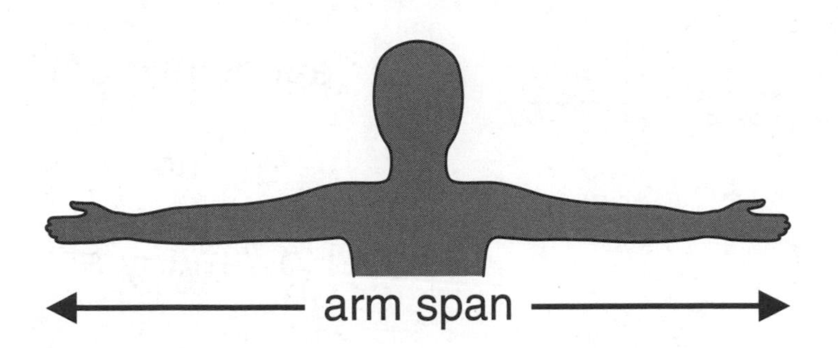

1. My arm span is ______ inches.

2. A middle value of arm spans for our class is ______ inches.

Math Boxes 7.7

1. Halve.

2 ____ 10 ____

4 ____ 14 ____

8 ____ 50 ____

2. Get 36 counters. Share them equally among 4 children.

How many counters does each child get?

____ counters

How many are left over?

____ counters

3. This is a ____-by-____ array.

● ● ● ●
● ● ● ●
● ● ● ●
● ● ● ●
● ● ● ●

How many dots in all?

____ dots

4. Find the rule. Complete the table.

Rule

in	out
193	183
232	222
441	
	346

5. Draw a rhombus. Make each side 2 cm long.

6. Solve.

Unit
cows

34 + ____ = 60

27 = 40 − ____

80 − 51 = ____

90 = 62 + ____

Date Time

Math Boxes 7.8

1. Write three different names for 28.

28 = ____ + ____ + ____

____ + ____ + ____ = 28

28 = ____ + ____ + ____

2. Make a ballpark estimate. Solve.

$$\begin{array}{r} 152 \\ -\ 129 \\ \hline \end{array}$$

Ballpark estimate:

$$\begin{array}{r} \square \\ -\ \square \\ \hline \square \end{array}$$

3. Write 4 different names for 50¢.

____

____

____

____

4. Match each person with the correct weight.

newborn baby	about 144 pounds
2nd grader	about 63 pounds
adult	about 7 pounds

5. 1 bag of sugar weighs 5 pounds.

6 bags of sugar weigh

____ pounds.

6. Dillon leaped 32 inches. Marcus leaped 27 inches. How many more inches did Dillon leap? ____ inches

Fill in the diagram.

Quantity	

Quantity	Difference

Date Time

The Lengths of Objects

Reminder: *in.* means *inches; cm* means *centimeters*

Measure each item to the nearest inch.
Measure each item to the nearest centimeter.
Record your answers in the blank spaces.

1. **pencil**

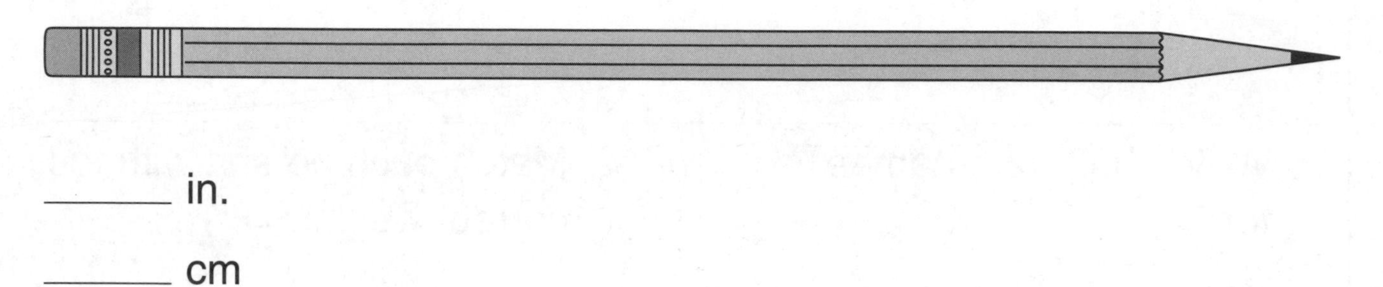

_______ in.

_______ cm

2. **screwdriver**

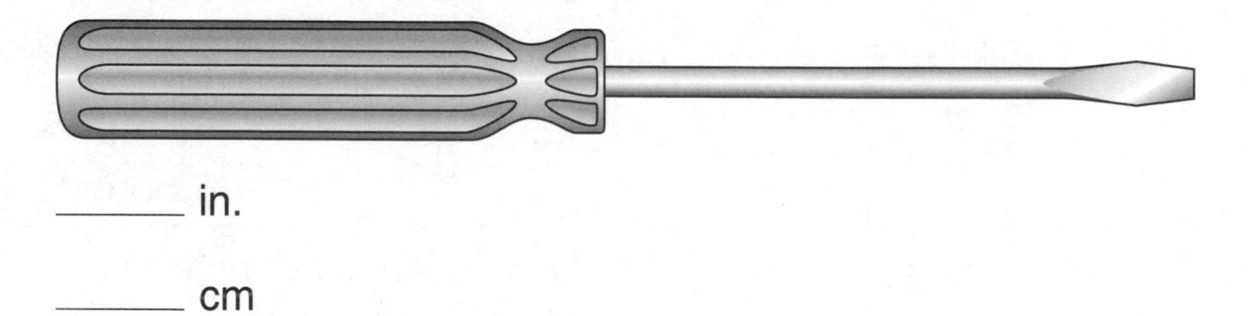

_______ in.

_______ cm

3. **pen**

_______ in.

_______ cm

Date Time

The Lengths of Objects (cont.)

4. bolt

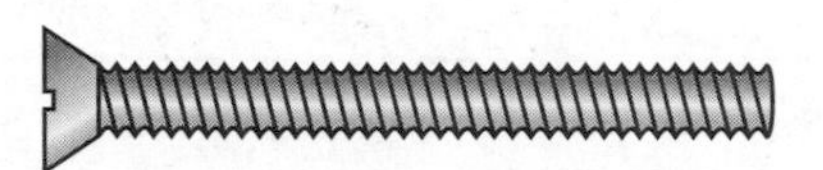

_______ in.

_______ cm

5. dandelion leaf

_______ in.

_______ cm

6. List the objects in order from shortest to longest.

____

____

____

____

____

Date Time

Table of Our Arm Spans

Make a table of the arm spans of your classmates.

Our Arm Spans

Arm Span (inches)	Frequency	
	Tallies	Number
	Total =	

Use with Lesson 7.9.

Date Time

Bar Graph of Our Arm Spans

Make a bar graph of the arm spans of your classmates.

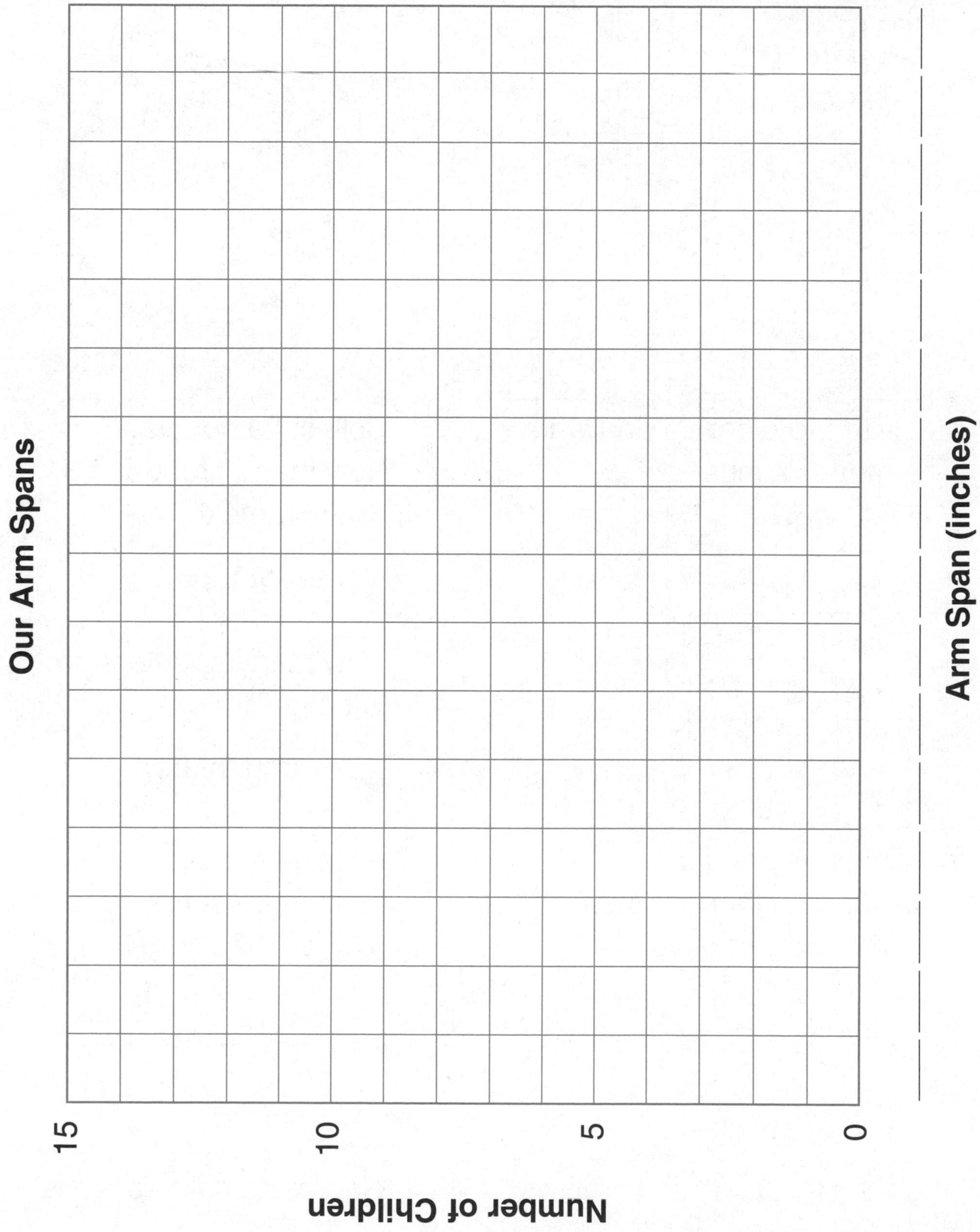

Our Arm Spans

15 10 5 0

Number of Children

Arm Span (inches)

Date Time

Math Boxes 7.9

1.

Rule
Double

in	out
2	
3	
	30
100	

2. Draw hands to show 4:40.

3. Fill in the diagram and write a model number.

Total	
Part	Part
12	18

____

4. Collect 29 counters. How many groups of 3 can you make?

______ groups

How many counters are left over?

______ counters

5. Use your calculator. Enter 42. Change to 70. Write what you did.

______

6. Arrange the allowances in order from smallest to largest. \$10, \$3, \$7, \$1, \$4

______, ______, ______,

______, ______

The median allowance is ______.

Date Time

Math Boxes 7.10

1. Count by 9s on the calculator. Start with 76.

76, ____, ____, ____, ____

What pattern do you see?

____

____

____

____

2. Use your calculator. Enter 27. Change to 60. Write what you did.

____

3.

Rule
+9

in	out
32	
56	
45	
	97
89	

4. Arrange the number of pets in order from smallest to largest. 7, 0, 4, 1, 3, 5, 2

____, ____, ____, ____,

____, ____, ____

The median number of pets is ____.

5. Write 5 names in the 90-box.

90

6. How many boxes are on this Math Boxes page?

____ boxes

How many boxes are on $\frac{1}{2}$ of this page?

____ boxes

Date Time

Equal Parts

Use a straightedge or Pattern-Block Template.

1. Divide the shape into 2 equal parts. Color 1 part.

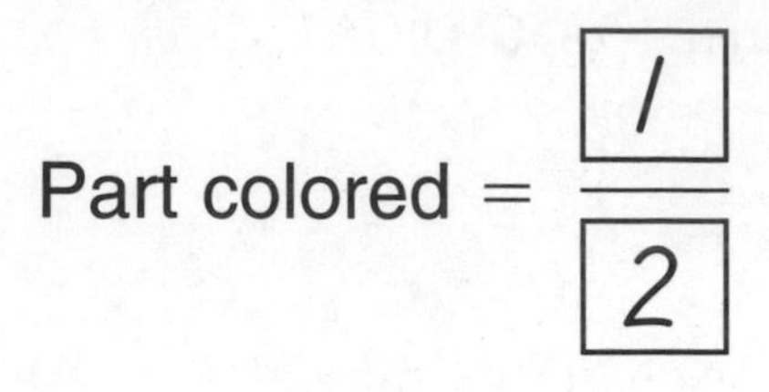

Part colored = $\frac{1}{2}$ Part not colored = $\frac{\square}{\square}$

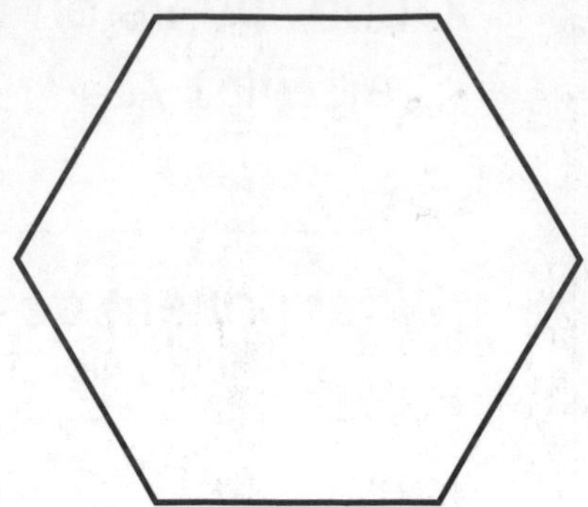

2. Divide the shape into 6 equal parts. Color 1 part.

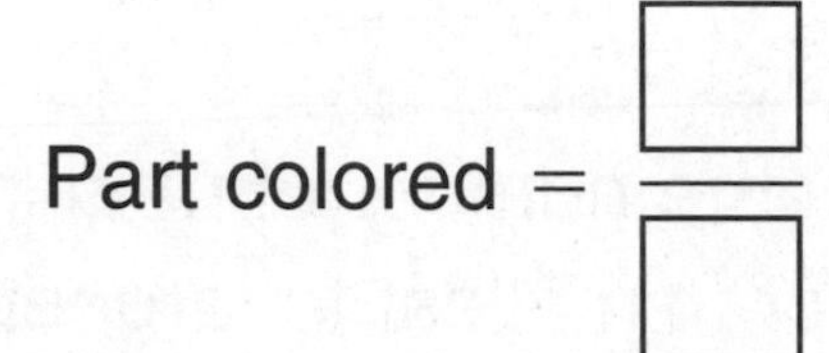

Part colored = $\frac{\square}{\square}$ Part not colored = $\frac{\square}{\square}$

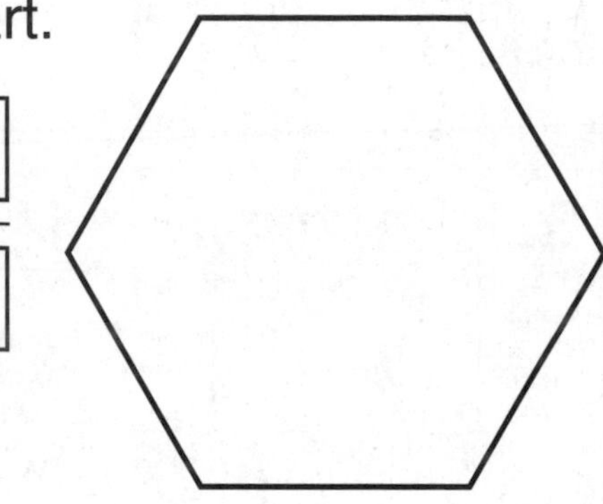

3. Divide the shape into 3 equal parts. Color 2 parts.

Part colored = $\frac{\square}{\square}$ Part not colored = $\frac{\square}{\square}$

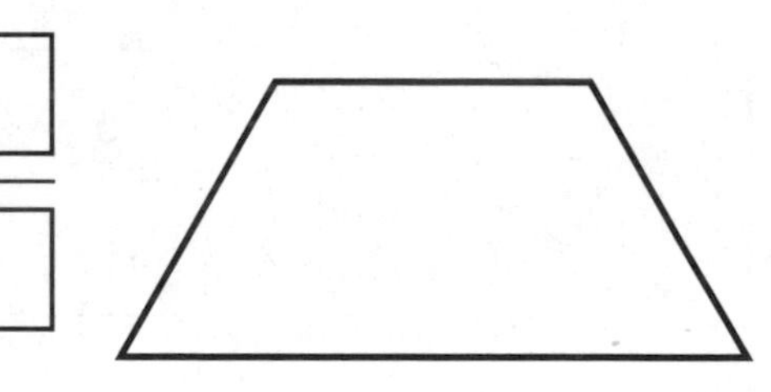

4. Divide the shape into 4 equal parts. Color 2 parts.

Part colored = $\frac{\square}{\square}$ Part not colored = $\frac{\square}{\square}$

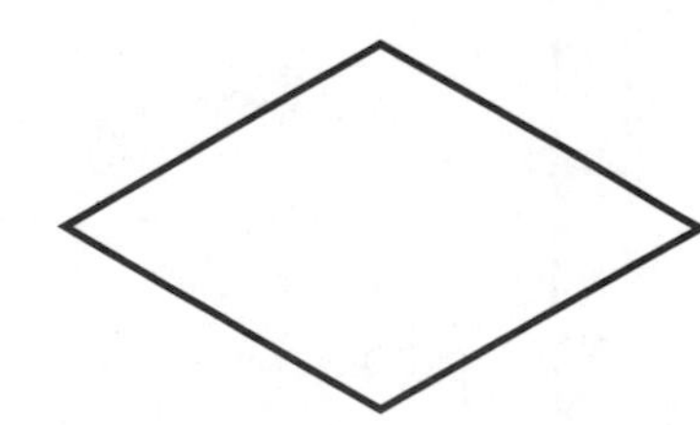

Date Time

Multiplication and Division Number Stories

Solve each problem. You can use the array to help.

In Problems 2 and 4, fill in the blanks with your own numbers.

1. The second graders want 24 oranges. There are 6 oranges in each bag. How many bags do they need?

They need ______ bags.

2. ______ bags of oranges

______ oranges per bag

How many oranges are there in all?

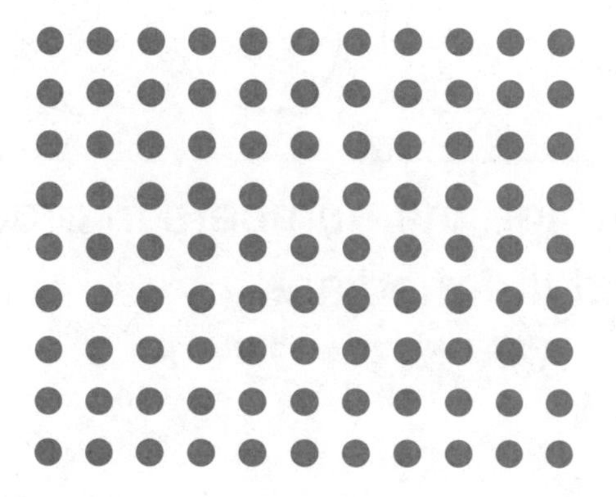

There are ______ oranges in all.

3. There are 30 kittens in all. Each mother cat had 5 kittens. How many mother cats are there?

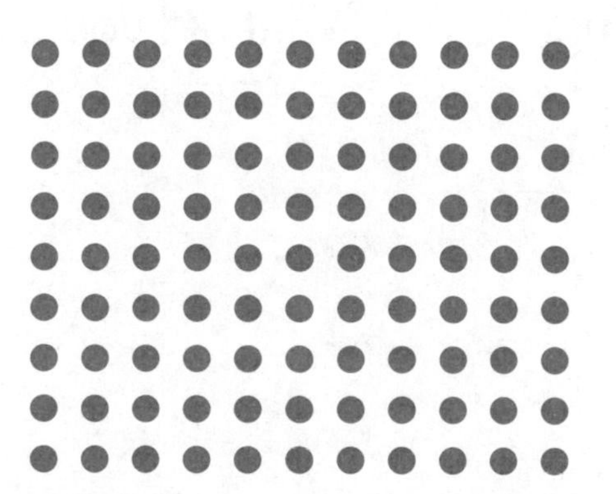

There are ______ mother cats.

4. ______ mother cats

______ kittens per mother cat

How many kittens are there in all?

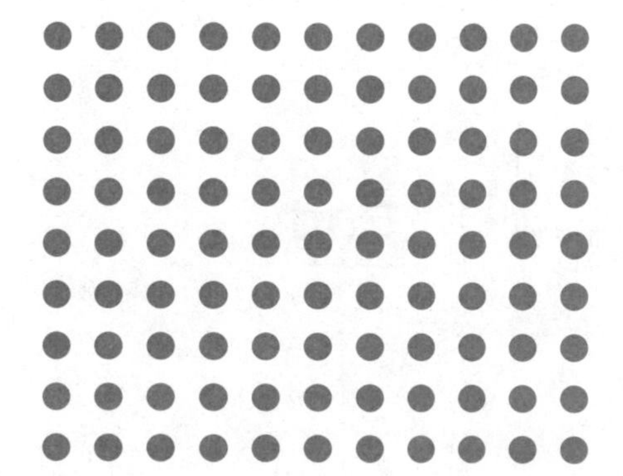

There are ______ kittens in all.

Math Boxes 8.1

1. Fill in the missing numbers.

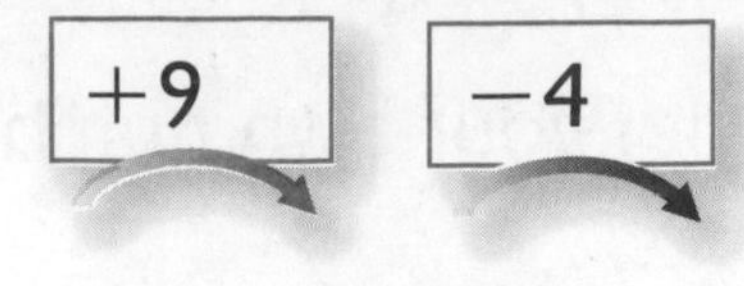

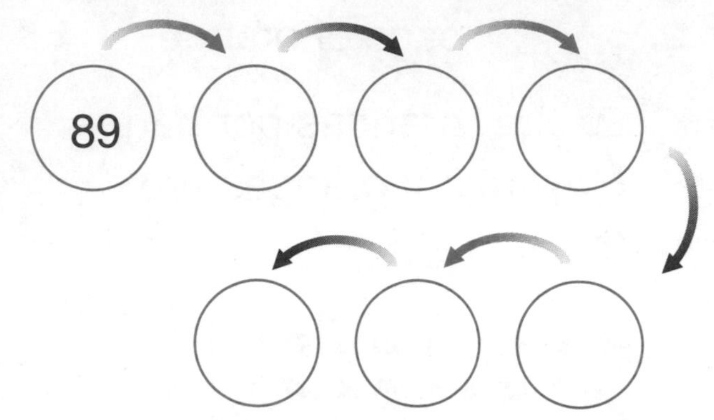

2. Write a number model for a ballpark estimate. Then subtract.

$$\begin{array}{r} 72 \\ -\ 56 \\ \hline \end{array}$$

Ballpark estimate:

____

3. Arrange the numbers in order. Find the median.
98 56 143 172 81

_____, _____, _____, _____,

_____

The median is _____.

4. In basketball you score 2 points for a basket and 1 point for each free throw. Katy scored seven 2-point baskets and two free throws. How many total points did Katy score?

_____ points

5. Solve the arrow-path puzzle.

Start

321

End

↓ → ↓ → ↑

6. Use your calculator. Enter 70. Change to 36. What did you do?

____

Date Time

Pattern-Block Fractions

Use pattern blocks to help you solve each problem.

Use your Pattern-Block Template to show what you did.

Example

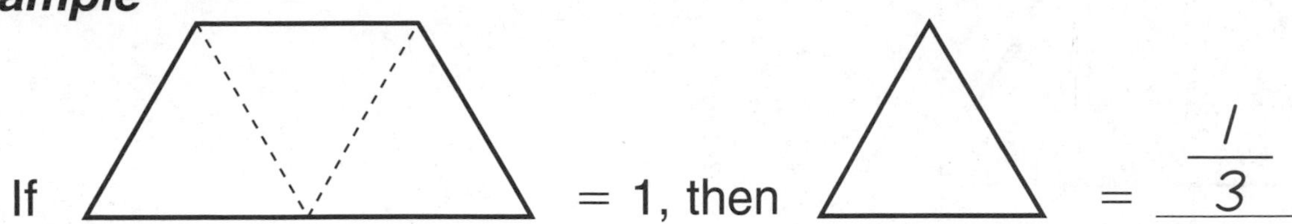

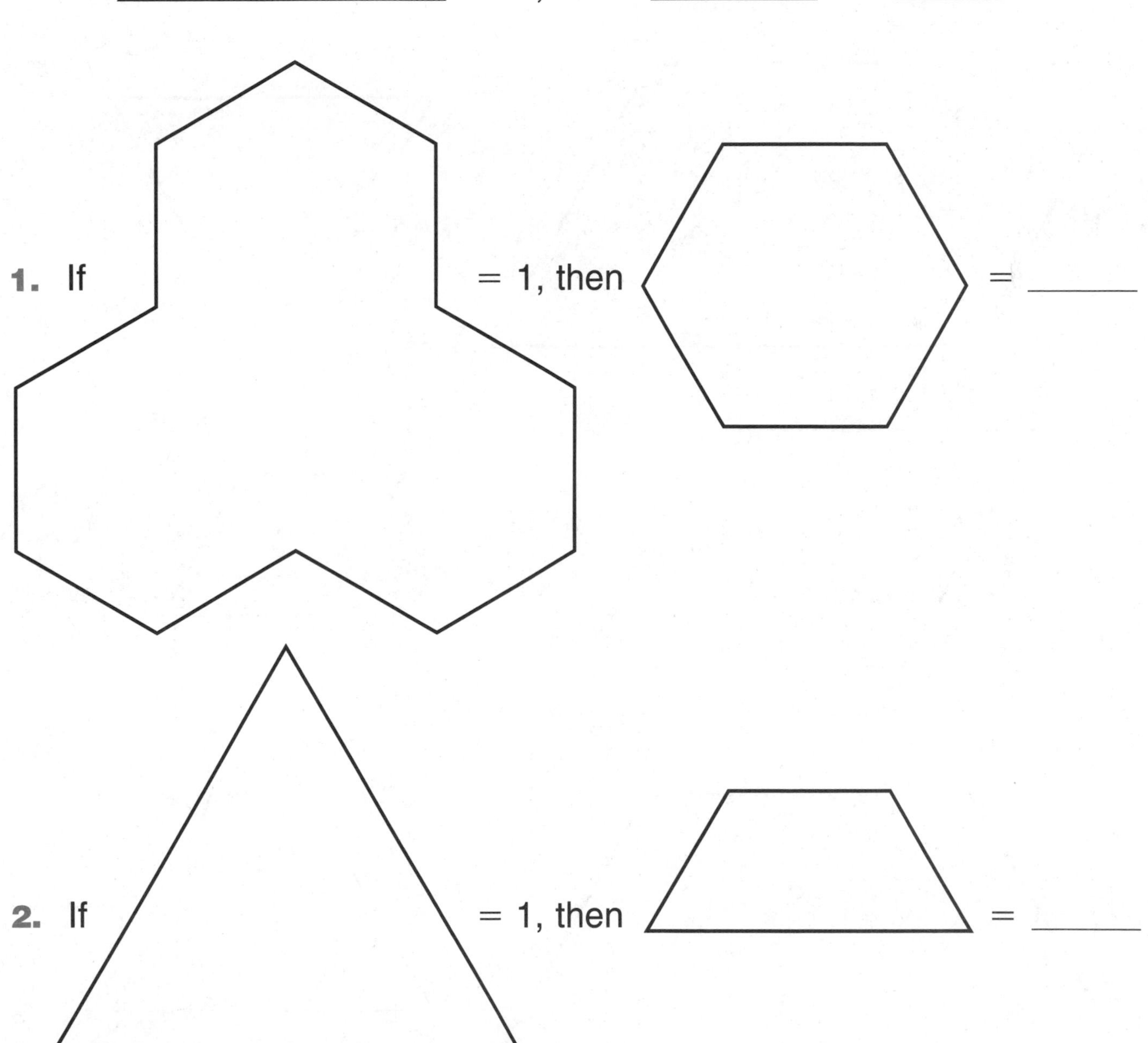

Date Time

Pattern-Block Fractions (cont.)

3. If 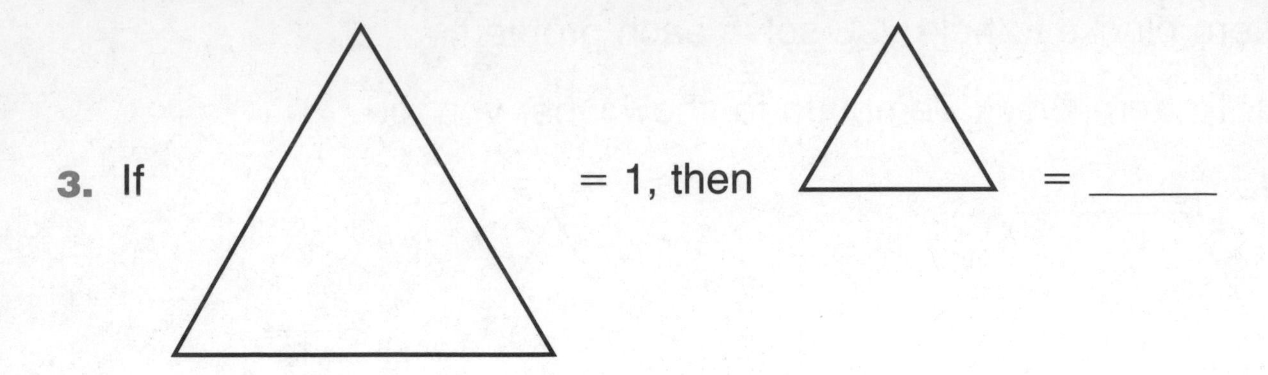 = 1, then = ______

4. If 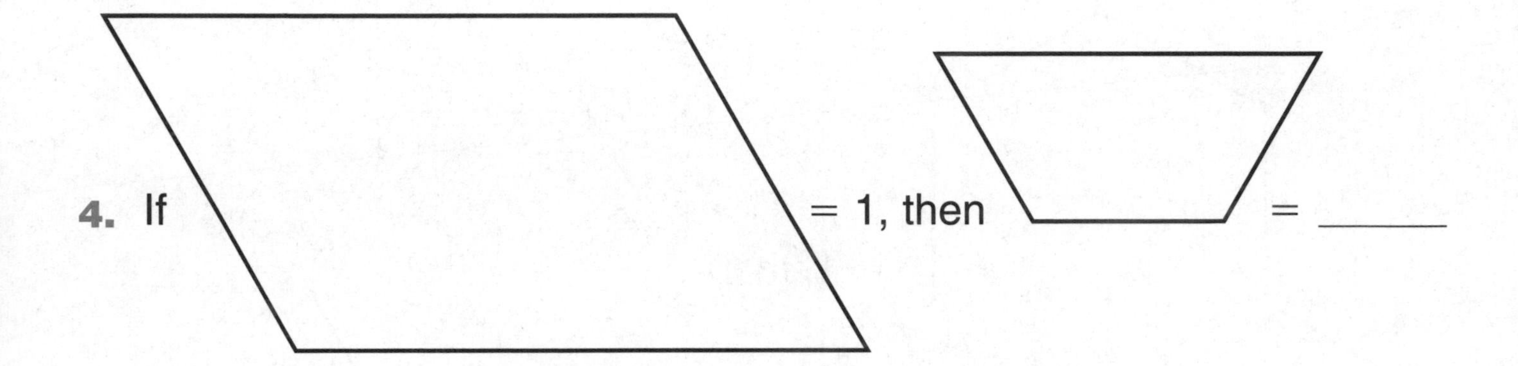 = 1, then = ______

5. If 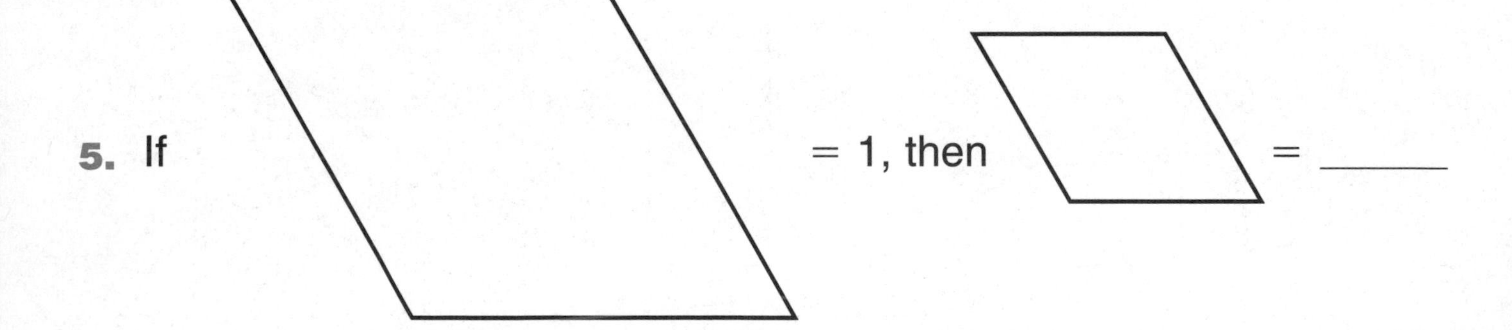 = 1, then = ______

6. If 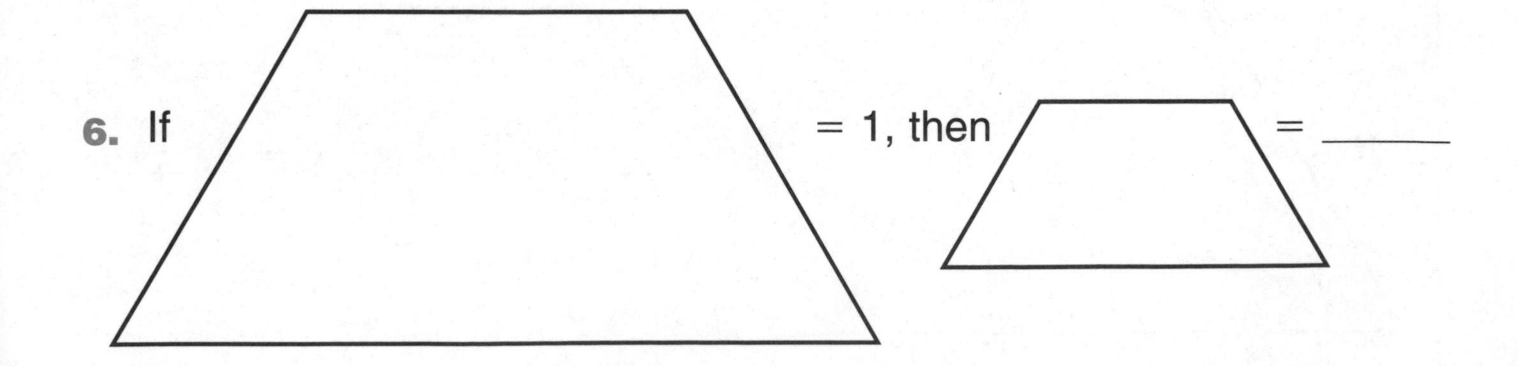 = 1, then = ______

Date Time

Geoboard Fences

1.

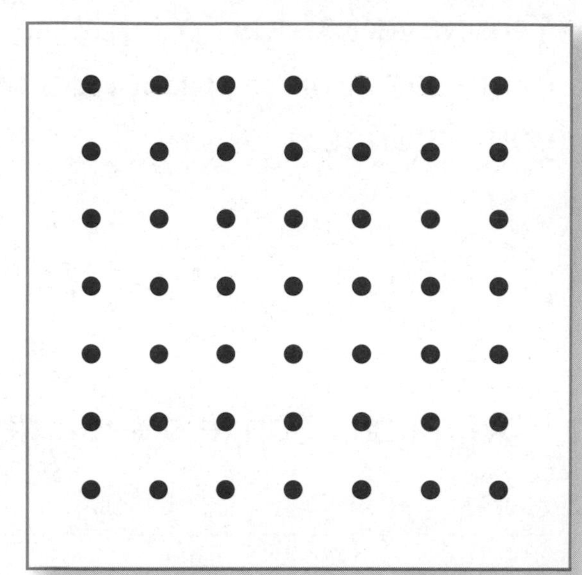

2.

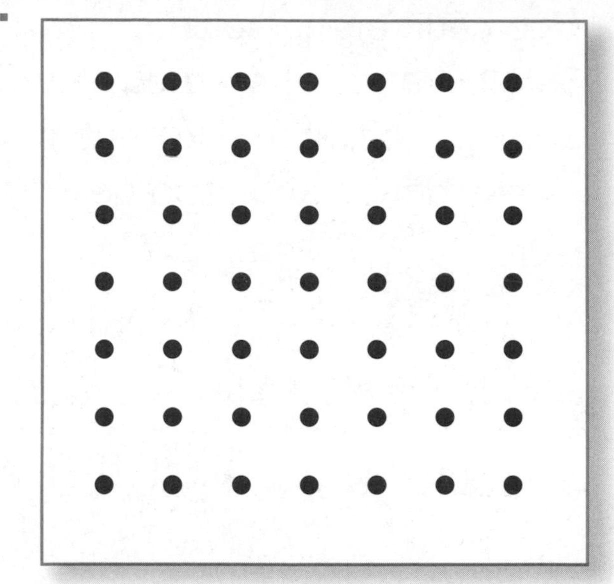

3.

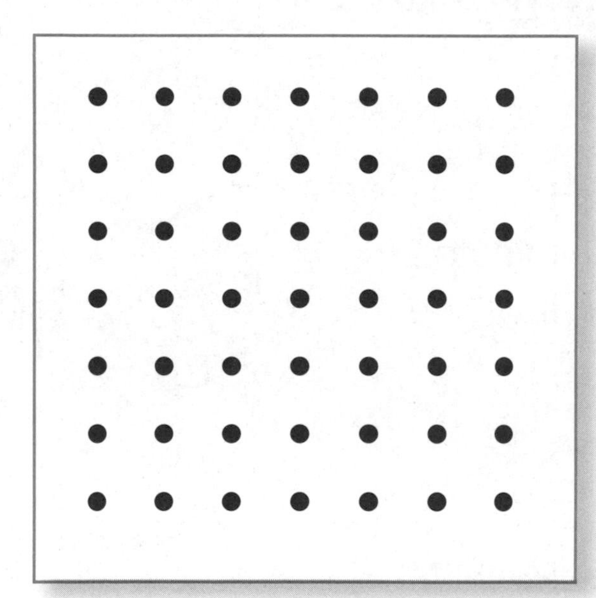

4.

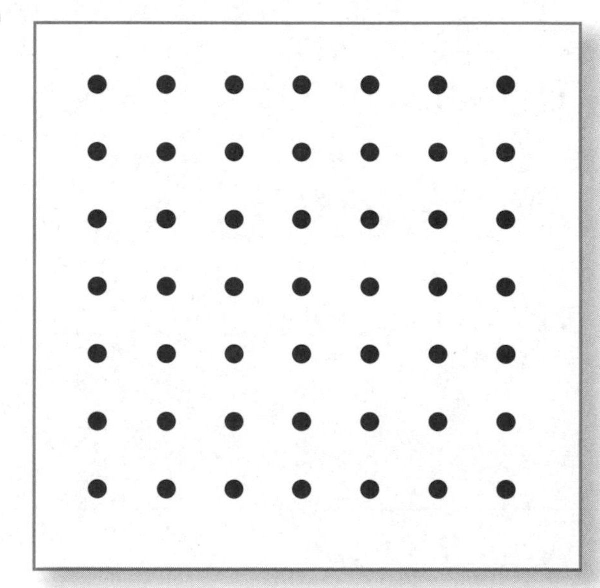

Fence	How many pegs in all?	How many rows of pegs?	How many in each row?
1.			
2.			
3.			
4.			

Date Time

Math Boxes 8.2

1. Use counters to solve.
35 blocks are shared equally among 3 children. How many blocks does each child get?

____ blocks

How many blocks are left over?

____ blocks

2. You have $1.00. You buy an ice cream cone for 63¢. Show your change.

How much change do you get?

____

3. 368

The value of 3 is ____________.

The value of 6 is ____________.

The value of 8 is ____________.

4. What time is it?

____ : ____

In 15 minutes, it will be

____ : ____.

5. Write fractions.

The part shaded = ____

The part not shaded = ____

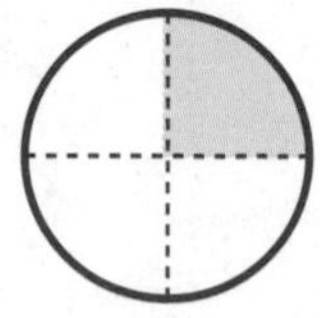

6. Double.

25¢ ______________

55¢ ______________

65¢ ______________

85¢ ______________

Date Time

Equal Shares

Use pennies to help you solve the problems.

Circle each person's share.

1. Two people share 10 pennies. How many pennies does each person get?

______ pennies

$\frac{1}{2}$ of 10 pennies = ______ pennies

2. Three people share 9 pennies. How many pennies does each person get?

______ pennies

$\frac{1}{3}$ of 9 pennies = ______ pennies

$\frac{2}{3}$ of 9 pennies = ______ pennies

3. Four people share 12 pennies. How many pennies does each person get?

______ pennies

$\frac{1}{4}$ of 12 pennies = ______ pennies

$\frac{3}{4}$ of 12 pennies = ______ pennies

Date Time

Fractions of Sets

A fraction is given in each problem. Color that fraction of the checkers red.

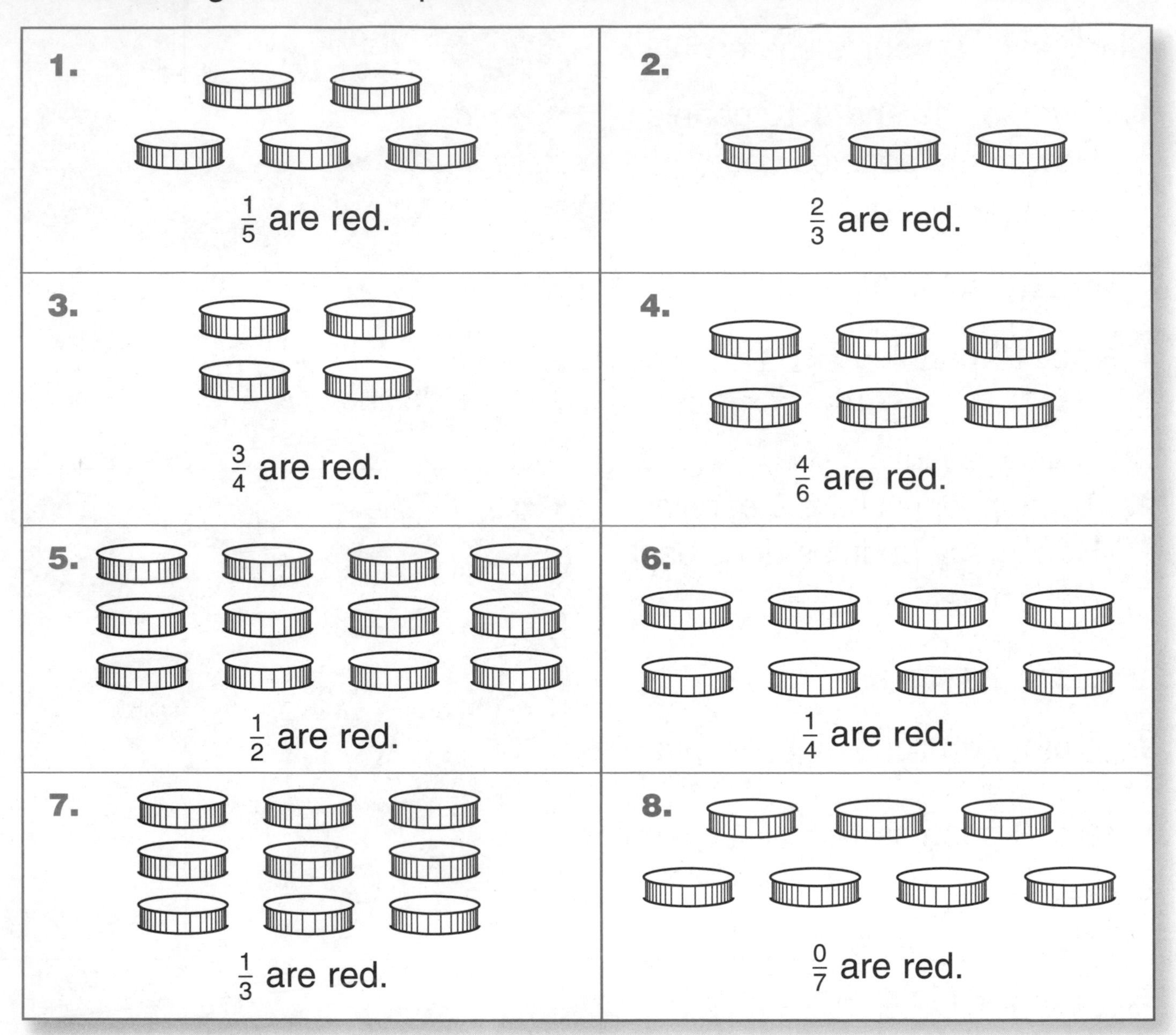

Challenge

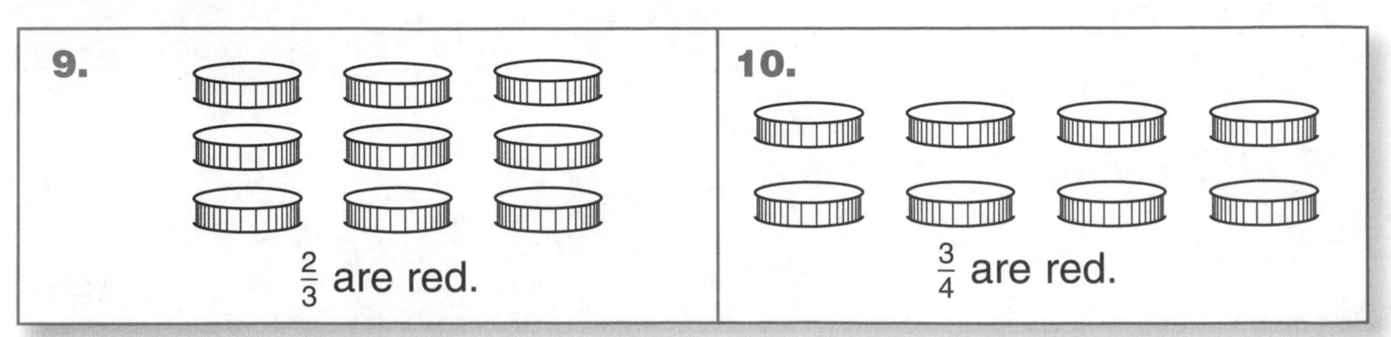

Date Time

Equal Parts

Use a straightedge or Pattern-Block Template.

1. Divide the shape into 4 equal parts. Color 3 parts.

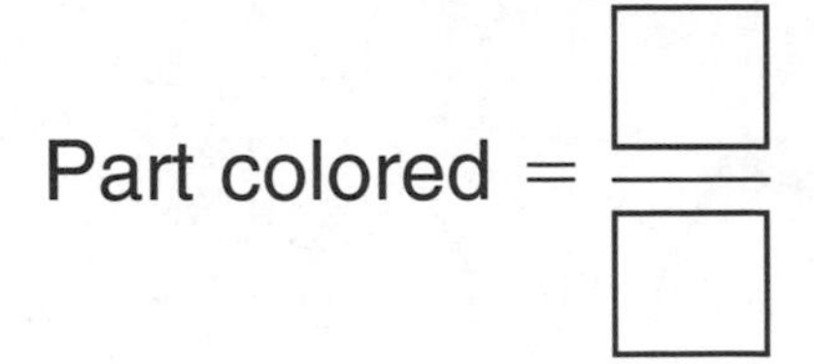

Part colored =

Part not colored =

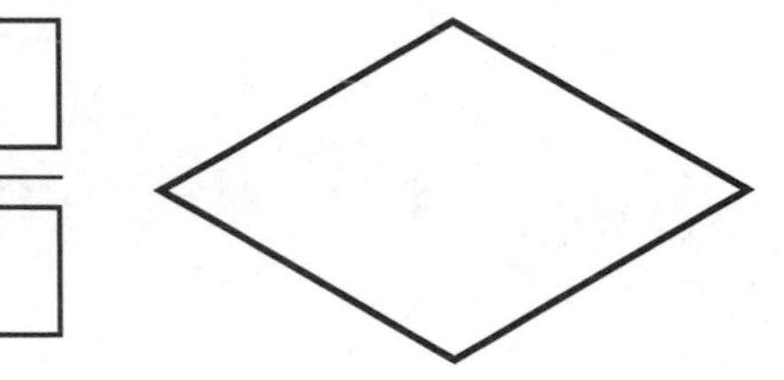

2. Divide the shape into 3 equal parts. Color 1 part.

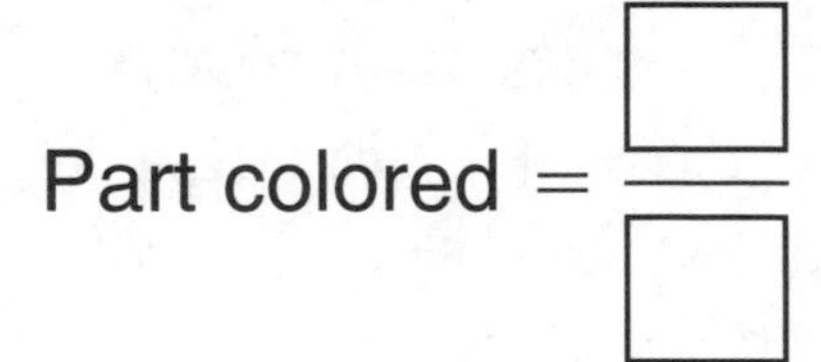

Part colored =

Part not colored =

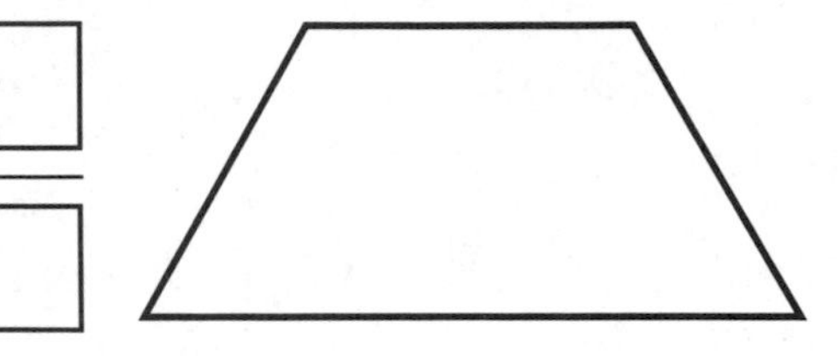

3. Divide the shape into 4 equal parts. Color 2 parts.

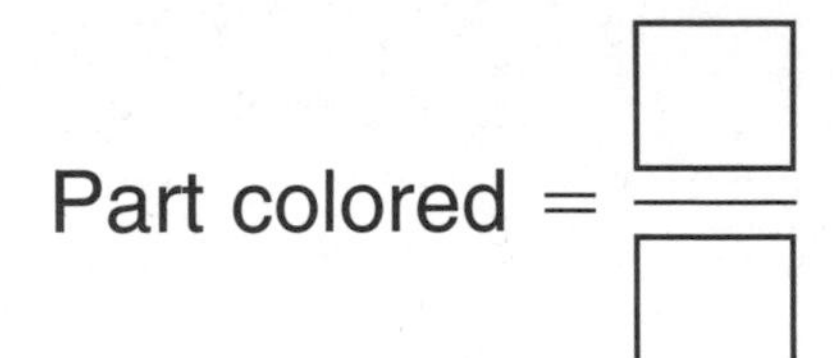

Part colored =

Part not colored =

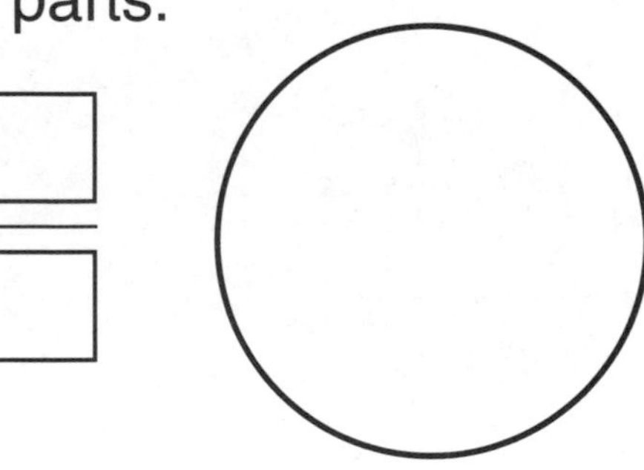

4. Divide the shape into 2 equal parts. Color 1 part.

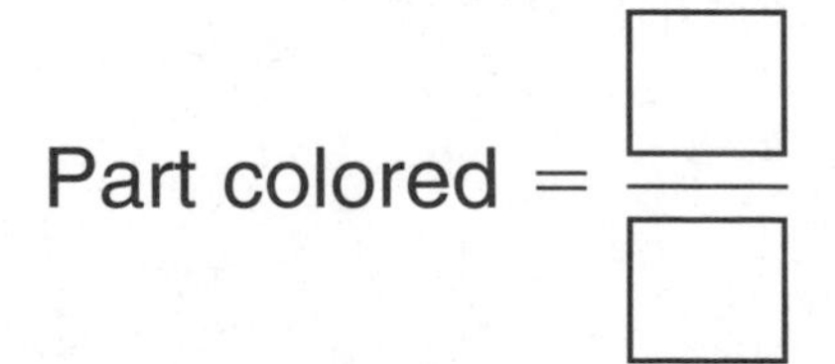

Part colored =

Part not colored =

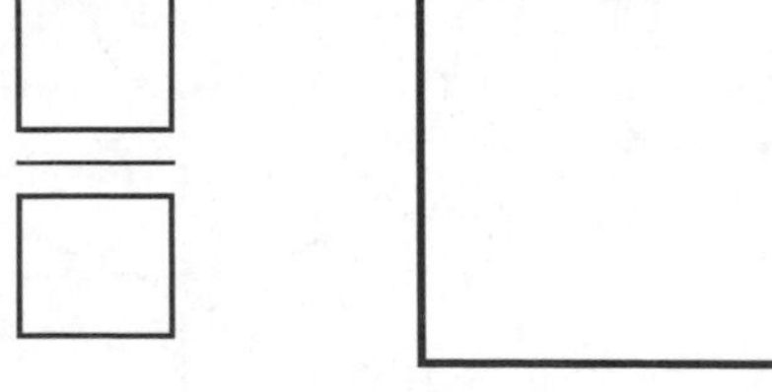

Date Time

Math Boxes 8.3

1. Write fractions.

The part shaded = ________

The part not shaded = ________

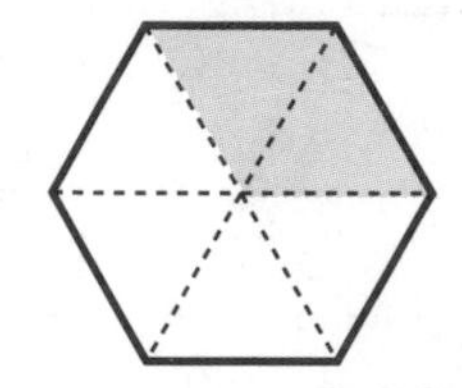

2. Circle the trapezoid that has $\frac{1}{3}$ shaded.

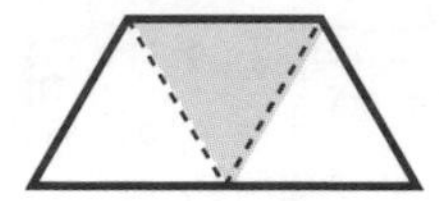

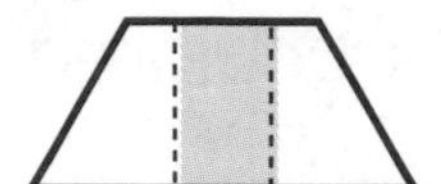

3. Solve.

Unit

_____ = 37 + 9

_____ = 137 + 9

116 − 8 = _____

176 − 8 = _____

4. Show $1.73 in two different ways. Use Ⓟ, Ⓝ, Ⓓ, and Ⓠ.

____

____

5. Complete.

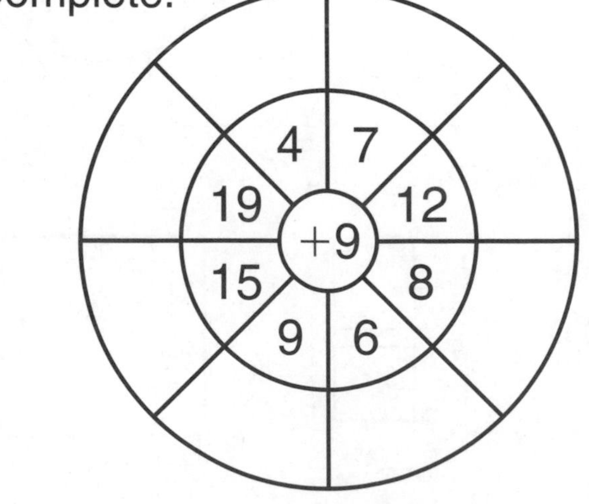

6. Make a 7 × 7 array with pennies.

How many pennies in all?

______ pennies

Math Boxes 8.4

1. If 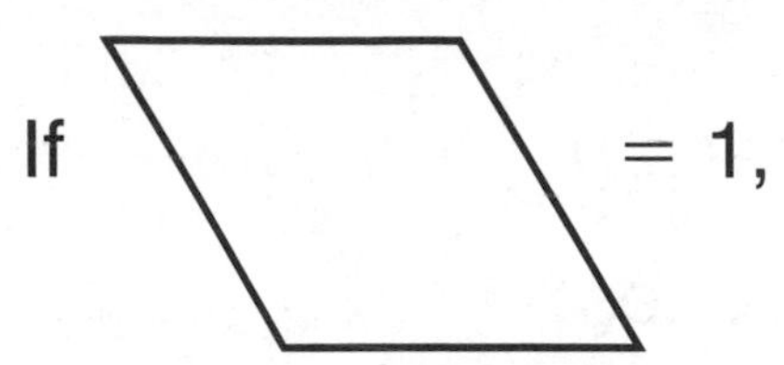 = 1,

then 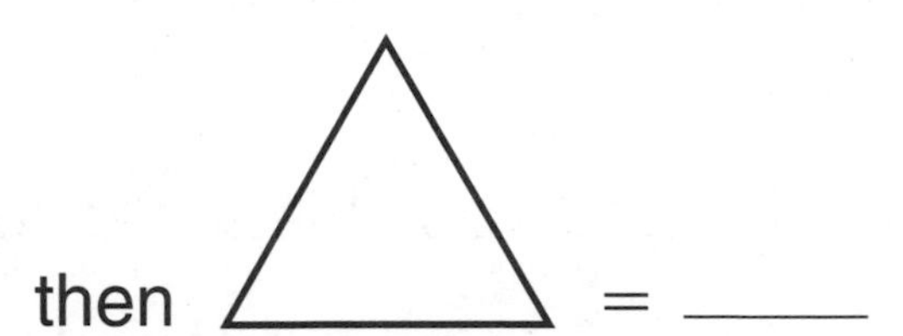 = ____

2. Color $\frac{5}{8}$ of the rectangle.

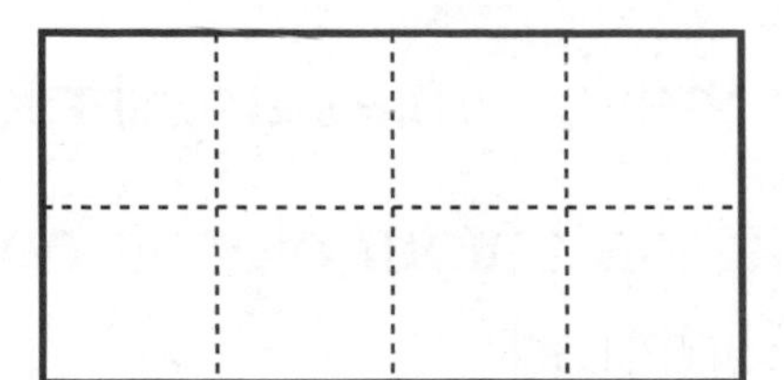

3. There are 9 dinosaurs.
3 are plant eaters.
Write a fraction to show how many are plant eaters.

____

4. Put the heights in order. Find the median height.
48 44 37 54 39

Unit
inches

____, ____, ____, ____,
____

The median height is
____ inches.

5. Complete the table.

Rule
$\frac{1}{2}$ of

in	out
2	1
4	
8	
	5

6. Draw an array with 5 fish bowls and 2 fish in each bowl.

How many fish in all?

____ fish

Equivalent Fractions

Do the following:

- Use the circles you cut out of *Math Masters,* page 144.
- Cut these circles apart along the dashed lines.
- Glue the cutout pieces onto the circles on this page and the next, as directed.
- Write the missing numerators to complete the equivalent fractions.

1. Cover $\frac{1}{2}$ of the circle with fourths.

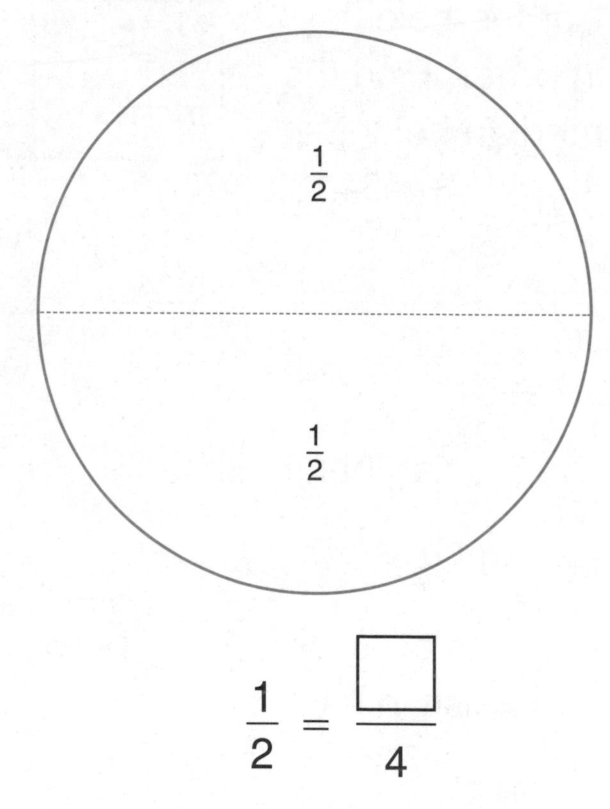

$\frac{1}{2} = \frac{\square}{4}$

2. Cover $\frac{1}{4}$ of the circle with eighths.

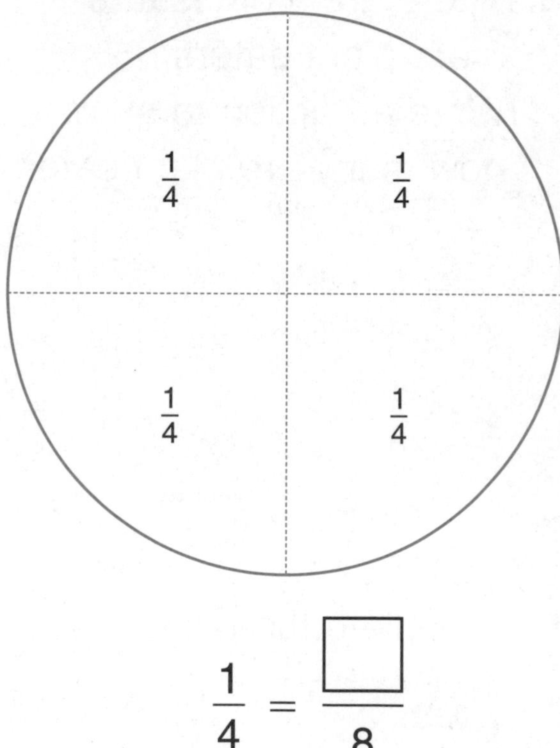

$\frac{1}{4} = \frac{\square}{8}$

Date Time

Equivalent Fractions (cont.)

3. Cover $\frac{2}{4}$ of the circle with eighths.

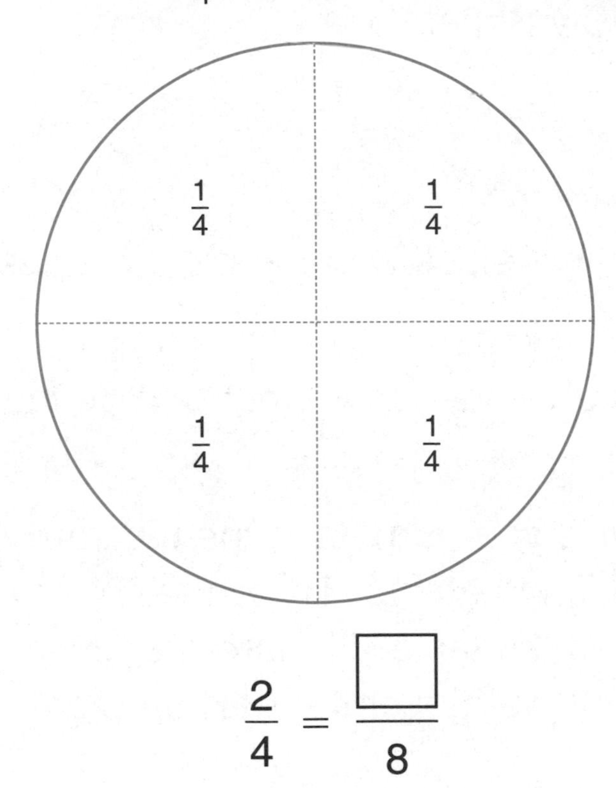

$\frac{2}{4} = \frac{\square}{8}$

4. Cover $\frac{1}{2}$ of the circle with sixths.

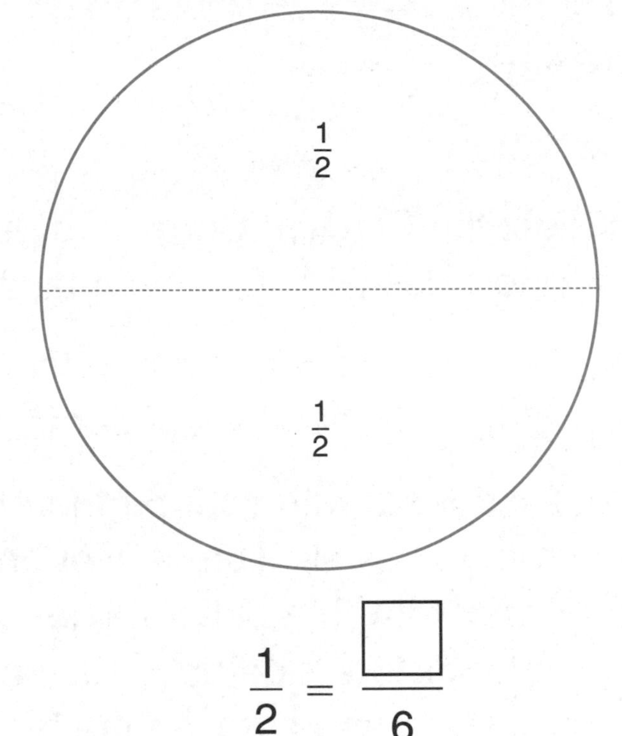

$\frac{1}{2} = \frac{\square}{6}$

5. Cover $\frac{1}{3}$ of the circle with sixths.

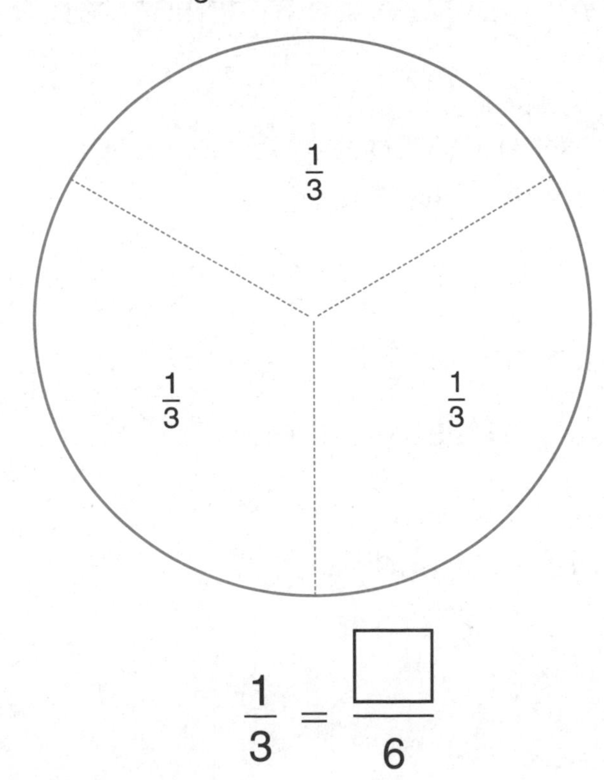

$\frac{1}{3} = \frac{\square}{6}$

6. Cover $\frac{2}{3}$ of the circle with sixths.

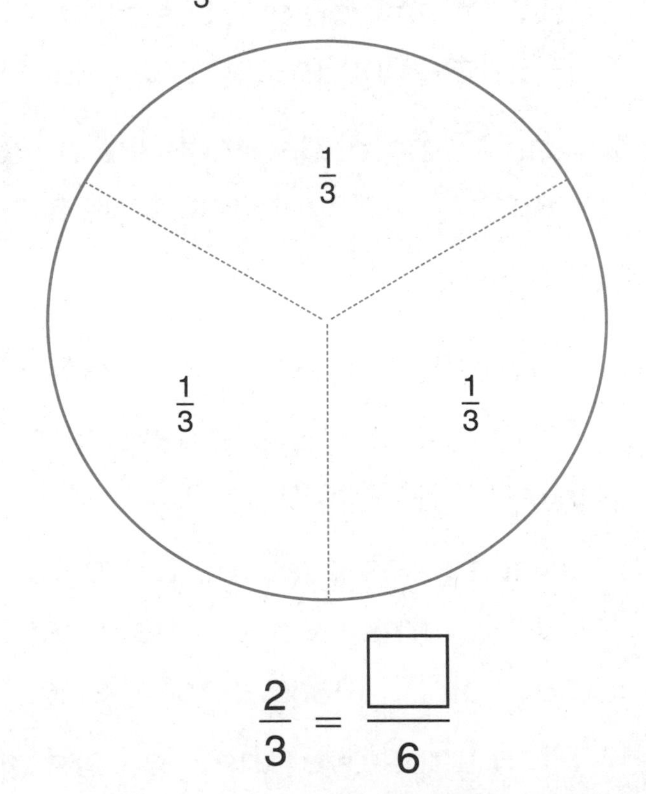

$\frac{2}{3} = \frac{\square}{6}$

Date Time

The *Equivalent Fractions Game*

Materials ❑ 32 Fraction Cards (2 sets cut from *Math Journal 2,* Activity Sheet 5)

Players 2

Directions

1. Mix the Fraction Cards and put them in a stack with the picture sides (the sides with the strips) facedown.
2. Turn the top card over so that the picture side faces up. Put it on the table near the stack of cards.
3. Take turns with your partner. When it is your turn, take the top card from the stack. Turn it over and put it on the table. Try to match this card with a picture-side-up card on the table. (If there are no other picture-side-up cards on the table, turn over the next card on the stack and put it on the table.)
4. Look for a match. If two cards match, take both of them. If there is a match that you don't see, the other player can take the matching cards. If there is no match, your turn is over.
5. The game ends when each card has been matched with another card. The player who took more cards wins the game.

Example

1. The top card is turned over. The picture shows $\frac{4}{6}$.
2. Ruth turns over the next card. It shows $\frac{2}{3}$. This matches $\frac{4}{6}$. Ruth takes both cards.
3. Justin turns over the top card on the stack. It shows $\frac{6}{8}$. Justin turns over the next card. It shows $\frac{0}{4}$. There is no match. Justin places $\frac{0}{4}$ next to $\frac{6}{8}$.
4. Ruth turns over the top card on the stack.

Date Time

The *Equivalent Fractions Game* (cont.)

Advanced Version

1. Mix the Fraction Cards and put them in a stack with the picture sides facedown.

2. Turn the top card over so that the picture side faces up. Put it on the table with the picture side faceup.

3. Players take turns. When it is your turn, take the top card from the stack, but *do not* turn it over. Keep the picture side down. Try to match the fraction on the card with one of the picture-side-up cards on the table.

4. If you find a match, turn your card over. Check that your match is correct by comparing the two pictures. If your match is correct, take both cards.

   If there is no match, place your card next to the other cards, picture side face-up. Your turn is over. If the other player can find a match, he or she can take the matching cards.

5. If there are no picture cards showing when Player 2 begins his or her turn, take the top card from the stack. Place it on the table with the picture side showing. Then Player 2 takes the next card in the stack and doesn't turn that card over.

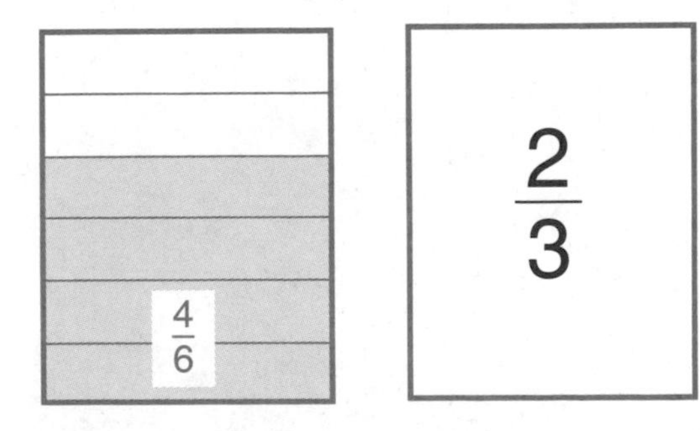

Marta thinks these two cards are a matching pair.

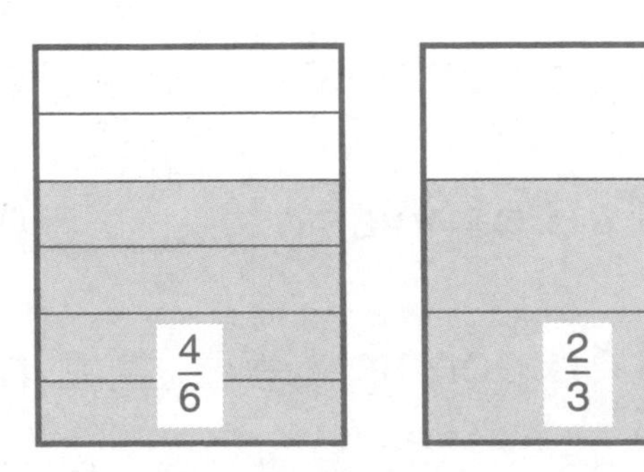

Marta checks that her match is correct by comparing the pictures of the fractions.

Date Time

Fractions of Collections

1. Five people share 15 pennies.

   How many pennies does each person get? ______ pennies

   $\frac{1}{5}$ of 15 pennies = ______ pennies

   $\frac{2}{5}$ of 15 pennies = ______ pennies

2. Six people share 12 pennies.

   How many pennies does each person get? ______ pennies

   $\frac{1}{6}$ of 12 pennies = ______ pennies

   $\frac{4}{6}$ of 12 pennies = ______ pennies

3. Four people share 16 pennies.

   How many pennies does each person get? ______ pennies

   $\frac{1}{4}$ of 16 pennies = ______ pennies

   $\frac{4}{4}$ of 16 pennies = ______ pennies

   $\frac{2}{4}$ of 16 pennies = ______ pennies

   $\frac{3}{4}$ of 16 pennies = ______ pennies

   $\frac{0}{4}$ of 16 pennies = ______ pennies

Date Time

Fractions of Collections (cont.)

Color the fractions of circles blue.

4. $\frac{3}{5}$ are blue.

5. $\frac{1}{2}$ are blue.

6. $\frac{1}{3}$ are blue.

7. $\frac{2}{3}$ are blue.

8. $\frac{3}{5}$ are blue.

9. $\frac{3}{8}$ are blue.

10. $\frac{3}{4}$ are blue.

11. $\frac{2}{6}$ are blue.

Date Time

Math Boxes 8.5

1. Circle $\frac{1}{5}$ of the nickels.

2. A pentagon has _____ sides.

A hexagon has _____ sides.

An octagon has _____ sides.

3. Use counters to solve.
13 marbles are shared equally. Each child gets 6 marbles. How many children are sharing?

_______ children

How many marbles are left over?

_____ marbles

4. Fill in the missing numbers.

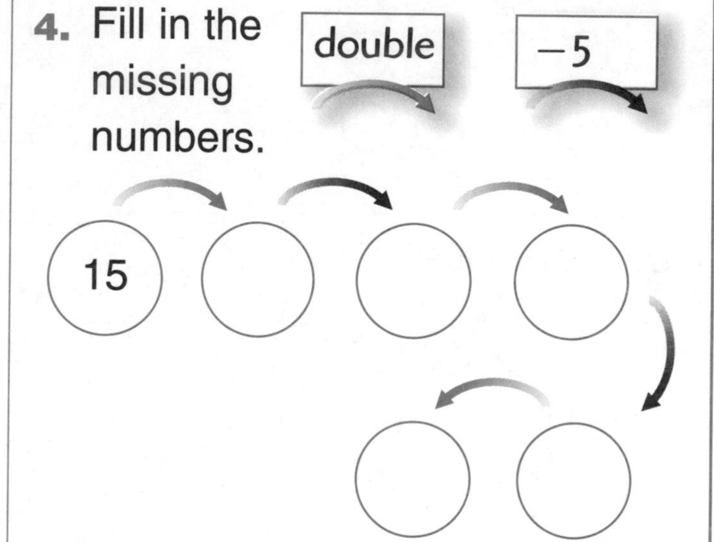

5. Solve the arrow-path puzzle.

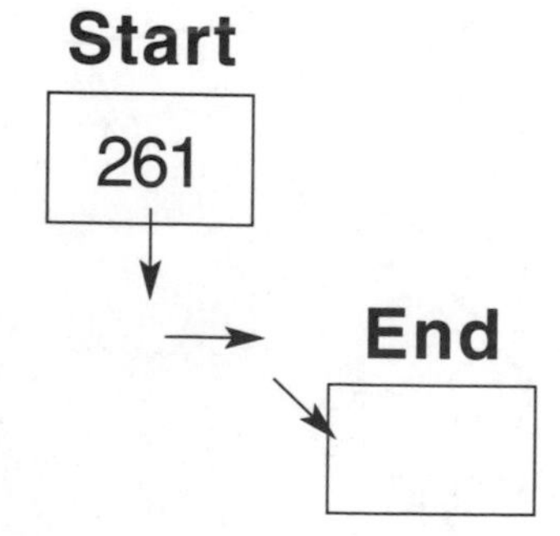

6. Draw a line segment 6 cm long. Divide the line segment into 3 equal parts.

Each part = _____ cm

Date Time

Math Boxes 8.6

1. Color $\frac{1}{2}$ of the set green.

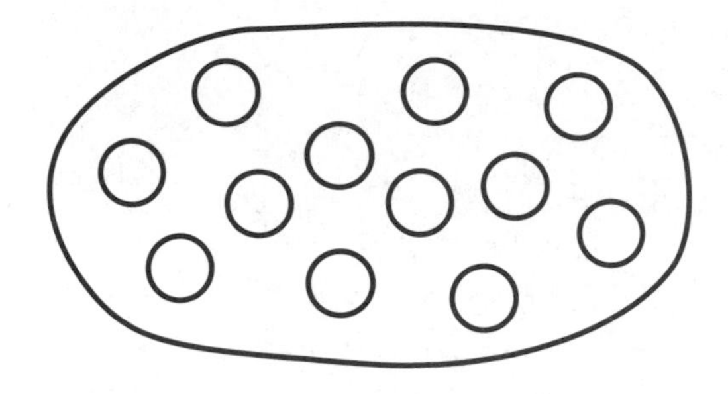

2. Complete the diagram.

Quantity	
64	
Quantity	**Difference**
28	

Write a number model.

____

3. Use the partial-sums algorithm to solve. Show your work.

$$\begin{array}{r} 78 \\ +\ 26 \\ \hline \end{array}$$

4. Circle the digits in the hundreds place.

1 2 8 9 7 2 4 6 3

2, 4 6 5 3, 0 9 1

6 6, 2 5 0

5. 10 balloons are shared equally among 3 children. How many balloons does each child get?

______ balloons

How many are left over?

______ balloons

6. Count 20 pennies.

$\frac{1}{2}$ = ______ pennies

$\frac{1}{4}$ = ______ pennies

$\frac{1}{5}$ = ______ pennies

Date Time

Halves—More or Less?

Use your Fraction Cards. List all the fractions that are:

less than $\frac{1}{2}$. ____________________

more than $\frac{1}{2}$. ____________________

the same as $\frac{1}{2}$. ____________________

Fraction Top-It

Materials ❑ 32 Fraction Cards (2 sets cut from *Math Journal 2*, Activity Sheet 5)

Players 2

Directions

1. Mix the Fraction Cards and put them in a stack so that all the picture sides (the sides with the strips) are facedown.
2. Each player turns over a card from the top of the stack. Players compare the shaded parts of their cards. The player with the larger (higher) fraction takes both cards.
3. If the shaded parts are equal, the fractions are equivalent. Each player turns over another card. The player with the larger fraction takes all the cards from both plays.
4. The game ends when all of the cards have been taken from the stack. The player who took more cards wins.

$\frac{1}{2}$ $\frac{1}{3}$

$\frac{1}{2}$ is greater than $\frac{1}{3}$.

Date Time

Fraction Top-It (cont.)

Advanced Version

1. Mix the Fraction Cards and put them in a stack so that all the picture sides (the sides with the strips) are facedown.
2. Each player takes a card from the top of the stack but does *not* turn it over.
3. Players take turns. When it is your turn, compare the fractions on the two cards. Say one of the following:
   - My fraction is more than your fraction.
   - My fraction is less than your fraction.
   - The fractions are equivalent.

John's card Barb's card

John says that his fraction is less than Barb's fraction.

4. Turn the cards over and compare the shaded parts. If you were correct, take both cards. If you were not correct, the other player takes both cards.

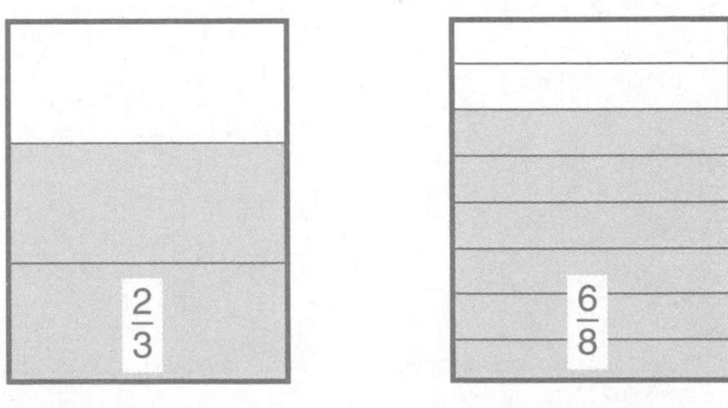

John's card Barb's card

Less of John's card is shaded: $\frac{2}{3}$ is less than $\frac{6}{8}$. John takes both cards.

Date Time

Fraction Number Stories

Solve these number stories. To help, you can use pennies or other counters, or you can draw pictures.

1. Mark has 4 shirts to wear.
   3 of them have short sleeves.
   What fraction of the shirts have short sleeves? ______

2. 8 birds are sitting on a tree branch.
   6 of the birds are sparrows.
   What fraction of the birds are sparrows? ______

3. June has 15 fish in her fish tank.
   $\frac{1}{3}$ of the fish are guppies.
   How many guppies does she have? ______

4. Mr. Sharp has 7 neckties.
   $\frac{3}{7}$ of the neckties are blue.
   How many neckties are blue? ______

5. Sam ate $\frac{0}{5}$ of a candy bar.
   How much of the candy bar did he eat? ________________

6. If you were thirsty, would you rather have $\frac{2}{2}$ of a carton of milk or $\frac{4}{4}$ of that same carton? Explain.

   ____

   ____

   ____

 Use with Lesson 8.7.

Date Time

Fraction Number Stories (cont.)

7. Joan and Terrell brought a 6-pack of soda to the picnic. They drank $\frac{2}{3}$ of the 6-pack. How many cans did they drink? ______

   What fraction of the 6-pack was left? ______

8. Write a fraction story. Ask your partner to solve it.

   ____

   ____

   ____

   ____

   ____

Challenge

9. Keesha's mother bought a dozen doughnuts. Keesha and her 2 brothers each ate 2 of the doughnuts for breakfast.

   How many doughnuts were left? ______

   What fraction of the doughnuts was left? ______

Date Time

Math Boxes 8.7

1. Karl ate $\frac{1}{2}$ of the pizza.

Nick ate $\frac{1}{4}$. Who ate more?

______

Draw a picture.

2. 6 children. Each has 4 stickers. How many stickers in all?

_____ stickers

Complete the multiplication diagram.

children	stickers per child	stickers in all

3. Complete.

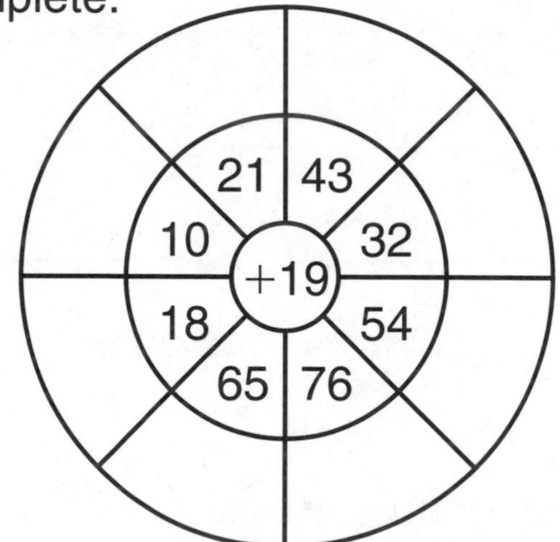

4. Color $\frac{1}{4}$ blue. Color $\frac{1}{4}$ yellow.

Color $\frac{1}{2}$ red.

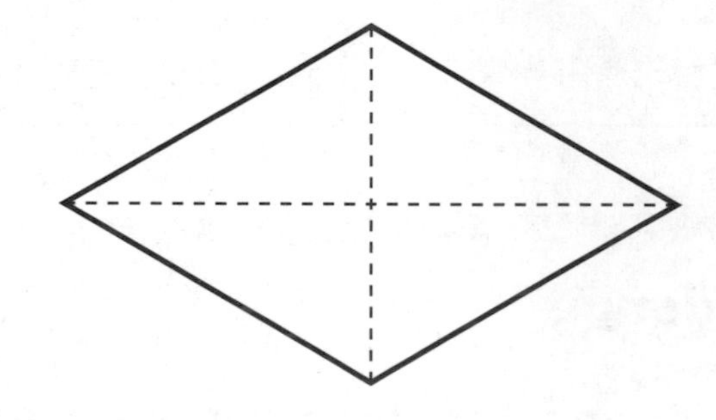

5. Write a fraction for each shaded part. Put >, <, or = in the box.

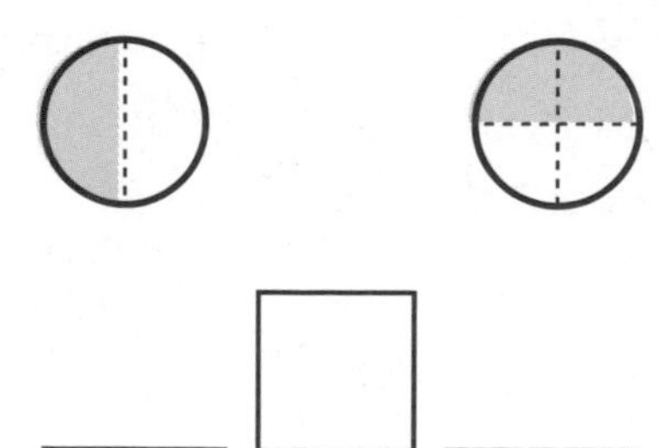

6. Draw hats on $\frac{1}{3}$ of the smiley faces.

Date Time

Math Boxes 8.8

1. 1 hour = _____ minutes

   $\frac{1}{2}$ hour = _____ minutes

   $\frac{1}{4}$ hour = _____ minutes

2. Write a fraction for each shaded part. Put $>$, $<$, or $=$ in the box.

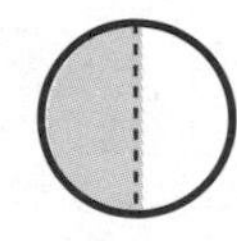

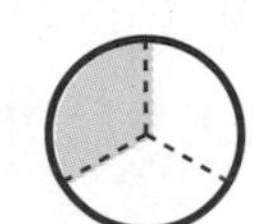

_____ ☐ _____

3. Draw a picture of 10 children.

   $\frac{1}{2}$ play ball. How many? _____

   $\frac{3}{10}$ jump rope. How many? _____

   $\frac{1}{5}$ skate. How many? _____

4. Draw a line segment 8 cm long. Now draw a line segment 5 cm shorter.

5. Draw a picture with 4 ladybugs and 5 spots on each ladybug.

   How many spots? _____

6. Fill in the missing amount.

   I had 38¢.

   I spent _____¢.

   I have 15¢ left.

Date Time

Yards

Materials ❑ yardstick

Directions

Record each step in the table below.

1. Choose a distance.
2. Estimate the distance in yards.
3. Use a yardstick to measure the distance to the nearest yard. Compare this measurement to your estimate.

Distance I Estimated and Measured	My Estimate	My Yardstick Measurement
	about ______ yards	about ______ yards
	about ______ yards	about ______ yards
	about ______ yards	about ______ yards
	about ______ yards	about ______ yards
	about ______ yards	about ______ yards

Date Time

Math Boxes 9.1

1. 15 children. $\frac{1}{3}$ are boys.

   How many are boys? _____

   How many are girls? _____

2. Divide the shape into 4 equal parts.

   What fraction of the shape is each part? _____

3. If 10¢ is ONE:

   what is 5¢? _____

   what is 20¢? _____

   If $1 is ONE:

   what is 25¢? _____

   what is $5? _____

4. Write 5 names for $\frac{1}{2}$. Use your Fraction Cards if you need help.

   $\frac{1}{2}$

5. Solve.

Unit

   72 − 12 = _____

   _____ = 72 − 28

   _____ = 72 − 34

   72 − 59 = _____

6. 9 cars. Each has 4 tires. How many tires in all?

cars	tires per car	tires in all

Date Time

Units of Linear Measure

Materials ❑ 12-inch ruler ❑ 10-centimeter ruler

Directions

1. Measure the length of two objects or distances.
2. First measure to the nearest foot. Measure again to the nearest inch.
3. Then measure to the nearest decimeter. Measure again to the nearest centimeter.

Object *or* Distance	Nearest Foot	Nearest Inch
	about _____ ft	about _____ in.
	about _____ ft	about _____ in.

Object *or* Distance	Nearest Decimeter	Nearest Centimeter
	about _____ dm	about _____ cm
	about _____ dm	about _____ cm

"What's My Rule?"

1. **Rule**: 1 ft = 12 in.

ft	in.
1	
2	
	36

2. **Rule**: 1 m = 100 cm

m	cm
1	
	300
10	

Date Time

Math Boxes 9.2

1. 17 pieces of gum are shared equally. Each child gets 4 pieces.

How many children are sharing? _____ children

How many pieces are left over? _____ pieces

2. Solve.

Unit
baby alligators

_____ = 24 + 41

33 + 12 = _____

_____ = 52 + 15

16 + 51 = _____

3. 246 228 273
209 298

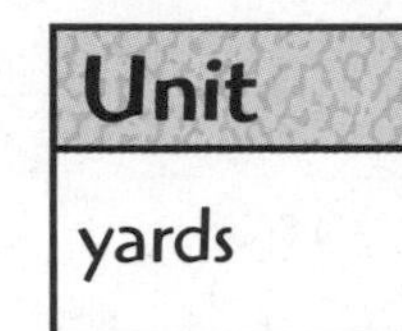

The median number of yards is _____.

4. Complete the diagram.
Then write a number model

Quantity
82

Quantity	Difference
39	

____

5. There are 6 rooms. Each room has 4 windows. How many windows in all?_____ windows
Draw an array.

6. Color $\frac{2}{8}$ red, $\frac{1}{2}$ yellow, and $\frac{1}{4}$ green.

Date Time

Measuring Lengths with a Ruler

Materials ❑ inch ruler

❑ centimeter ruler

Directions

Work with a partner. Use your ruler to measure the length of each object to the nearest $\frac{1}{2}$ inch and $\frac{1}{2}$ centimeter.

1. **craft stick**

about ______ inches long about ______ centimeters long

2. **large paper clip**

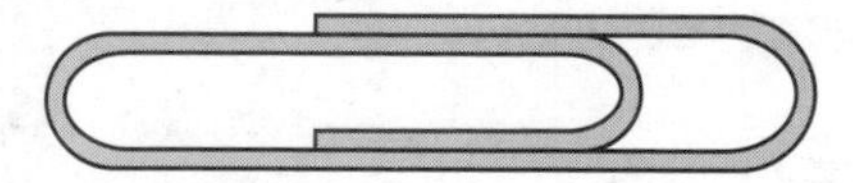

about ______ inches long about ______ centimeters long

3. **small paper clip**

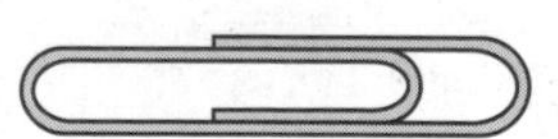

about ______ inches long about ______ centimeters long

4. **pencil**

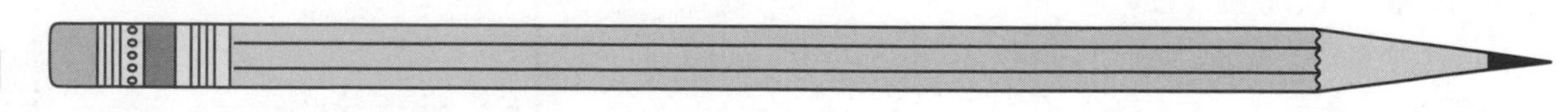

about ______ inches long about ______ centimeters long

Date Time

Measuring Lengths with a Ruler (cont.)

5. pencil 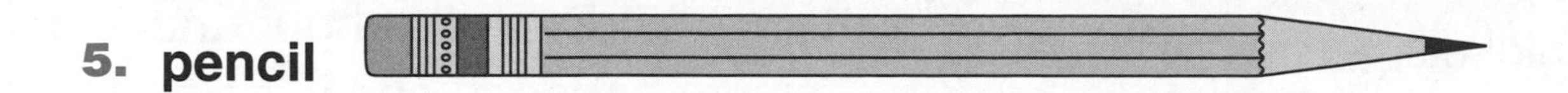

about ______ inches long about ______ centimeters long

6. nail 

about ______ inches long about ______ centimeters long

7. chalk

about ______ inches long about ______ centimeters long

8. pen

about ______ inches long about ______ centimeters long

9. straw

about ______ inches long about ______ centimeters long

Math Boxes 9.3

1. 2 nickels = 1 dime

1 nickel = $\frac{1}{2}$ dime

1 inch = _____ foot

_____ nickels = 1 quarter

1 nickel = _____ quarter

2. It is 8:45. Draw the hour and minute hands to show the time 15 minutes later. What time does the clock show?

_____:_____

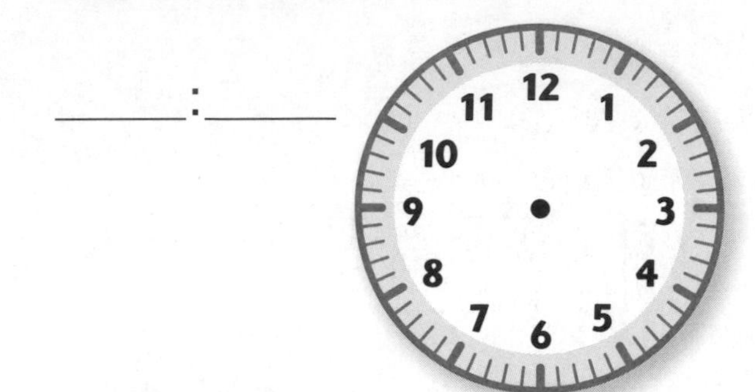

3. How many dots are in this 5-by-5 array?

_____ dots in all

4.

Rule
12 in. = 1 ft

in.	ft
6	
	2
48	

5. Circle the unit that makes sense.

A table is about

36 _____ long. in. yd

A school bus is about

18 _____ long. cm m

A newborn baby is about

20 _____ long. in. cm

6. Circle $\frac{5}{8}$.

Date Time

Distance Around and Perimeter

Measure the distance around the following to the nearest centimeter.

1. Your neck: ______cm 2. Your ankle: ______ cm

Measure the distance around two other objects to the nearest centimeter.

3. Object: ______________________________ Measurement: ______ cm

4. Object: ______________________________ Measurement: ______ cm

Measure the sides of each figure to the nearest $\frac{1}{2}$ inch.
Write the length next to each side. Then find the perimeter.

5. Perimeter: __________ inches

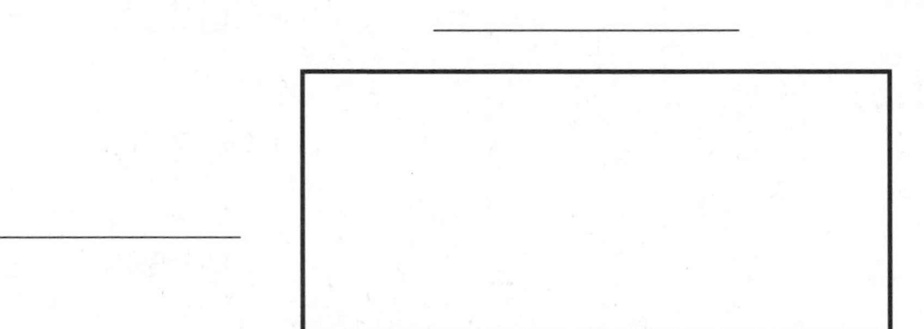

6. Perimeter: __________ inches

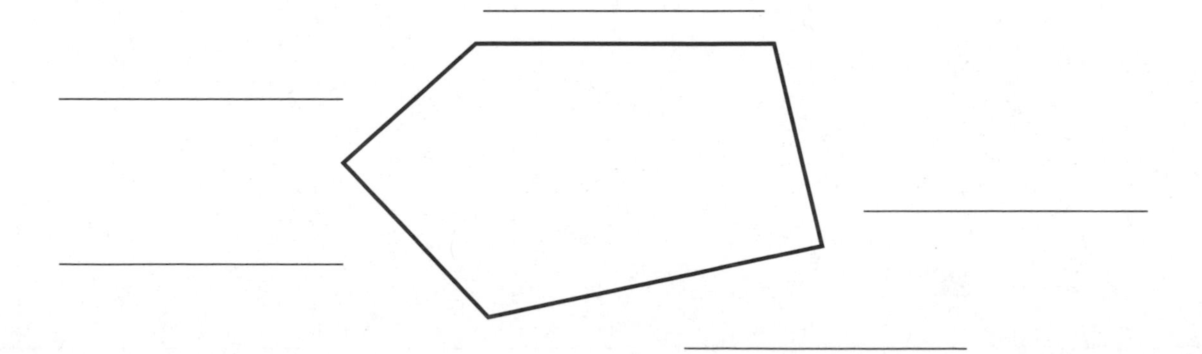

Date Time

Math Boxes 9.4

1. Draw a line segment $3\frac{1}{2}$ inches long.

Now draw a line segment 1 inch shorter.

2. Complete the frames.

Rules: +10, $\frac{1}{2}$

90 → ☐ → ☐ → ☐ → ☐ → 40

3.

Rule
10 dm = 1 m

dm	m
5	
40	
	6

4. Get 21 counters.

$\frac{1}{3}$ = _____ counters

$\frac{2}{7}$ = _____ counters

$\frac{3}{3}$ = _____ counters

5. Solve.

Unit

86 − 40 = _____

_____ = 198 − 60

259 − 40 = _____

_____ = 243 − 120

6. Count by 100s.

_______; _______; 2,748;

_______; _______; _______;

_______; _______

Date Time

Driving in the West

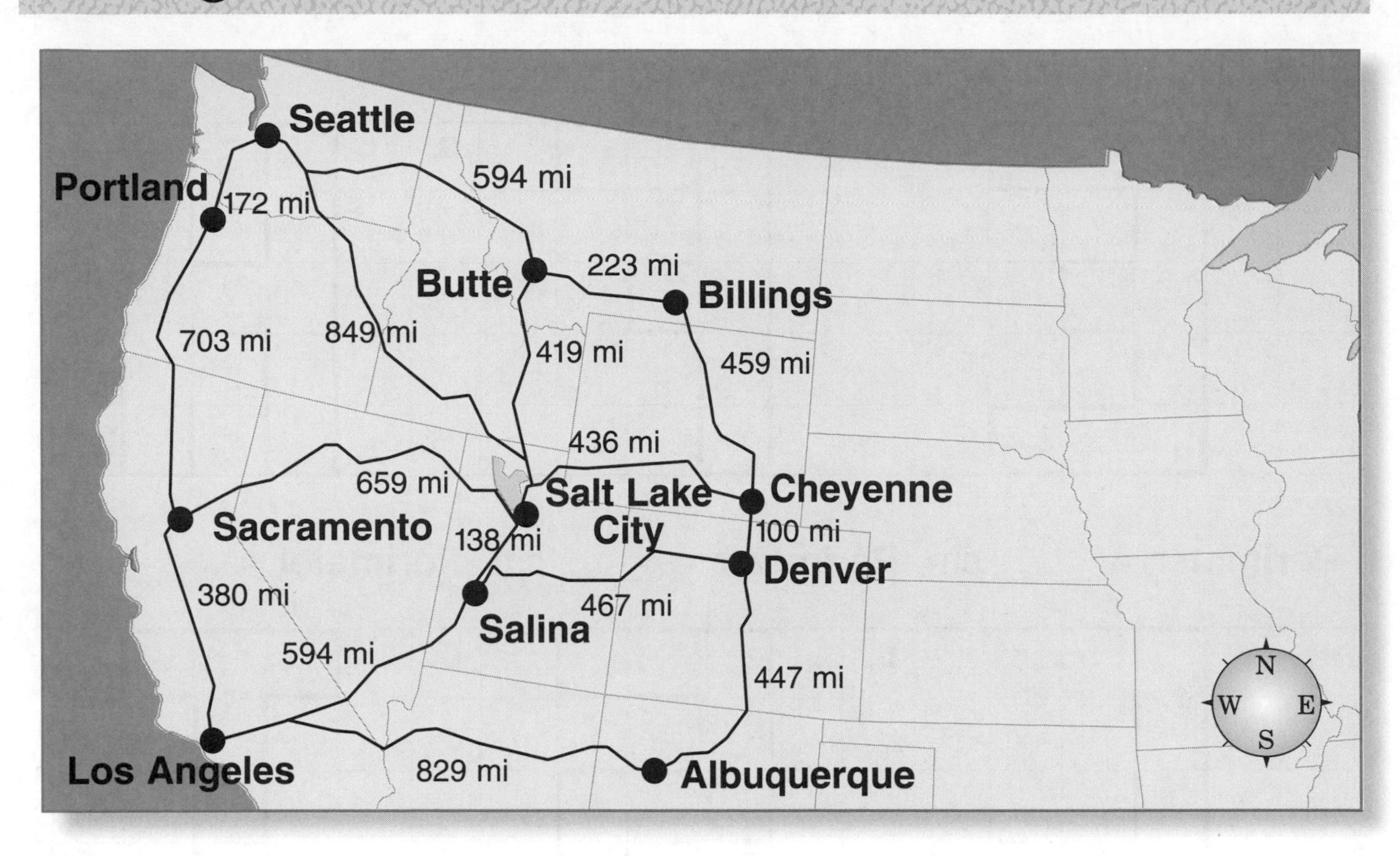

1. What is the shortest route from Seattle to Albuquerque?

____

____

2. Put a check mark in front of the longer trip.

_____ Salt Lake City to Billings by way of Butte

_____ Salt Lake City to Billings by way of Cheyenne

How much longer is that trip? about _____ miles longer

Date Time

Letter Perimeters

Find the perimeter of each letter.

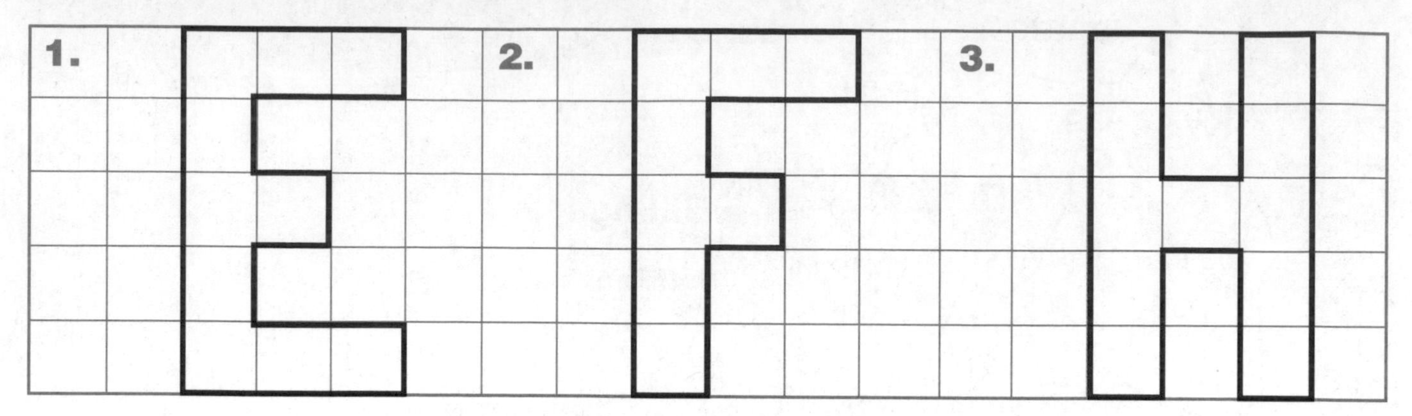

Perimeter = ____ cm **Perimeter** = ____ cm **Perimeter** = ____ cm

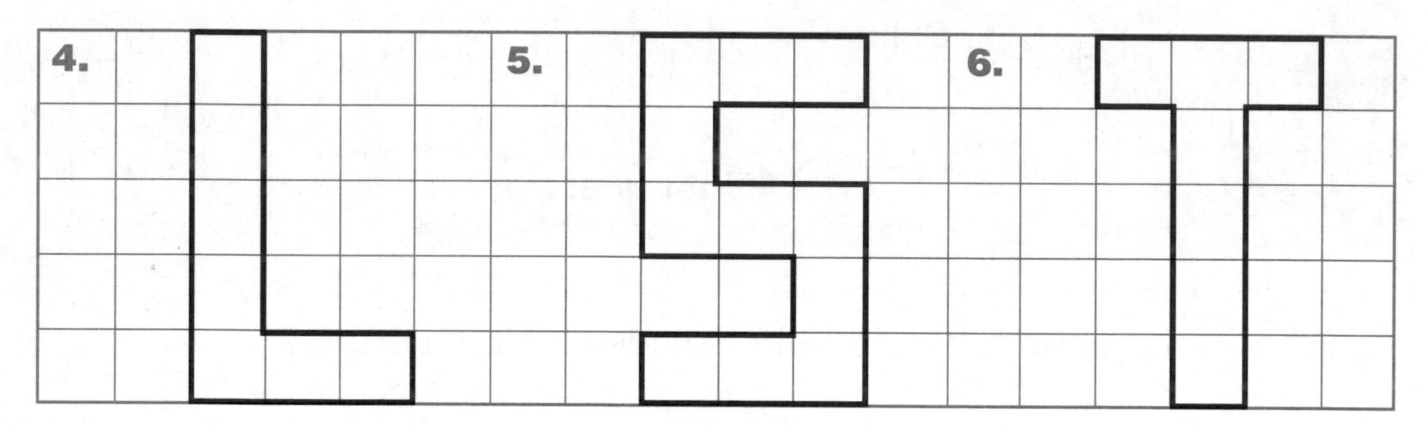

Perimeter = ____ cm **Perimeter** = ____ cm **Perimeter** = ____ cm

For Problems 7 and 8, draw a polygon and find the perimeter.

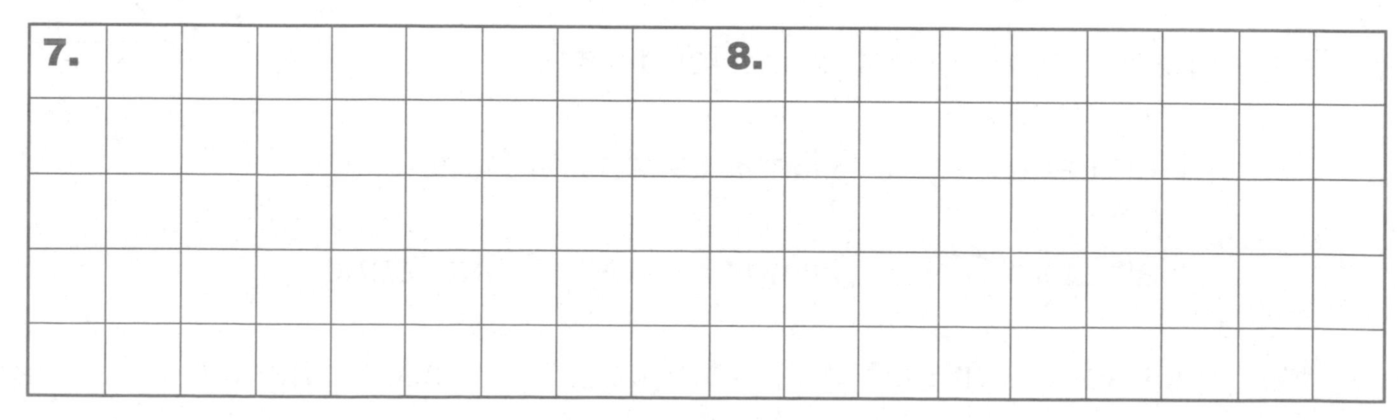

Perimeter = ____ cm **Perimeter** = ____ cm

Date Time

Math Boxes 9.5

1. Measure each side of the triangle to the nearest inch. Find the perimeter.

The perimeter is

_____ inches.

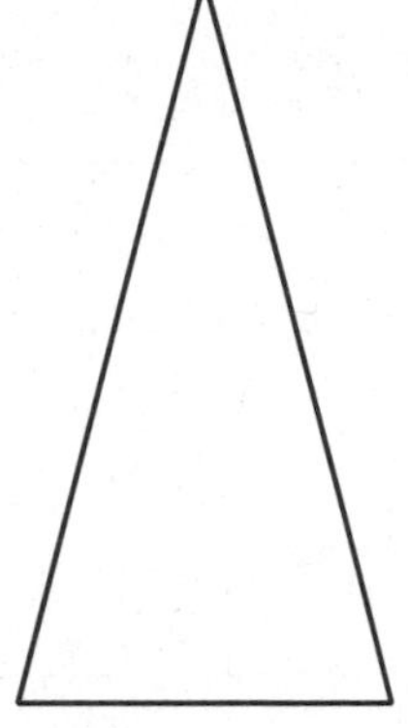

2. Ginny had 16 baseball cards. She traded $\frac{1}{4}$ of them.

How many cards did she trade?

_____ cards

How many did she keep?

_____ cards

3. Draw a line segment $4\frac{1}{2}$ cm long.

Now draw a line segment 2 cm longer.

4. If 25¢ is ONE:

what is 5¢? _____

what is 50¢? _____

If $2 is ONE:

what is 50¢? _____

what is $4? _____

5.

Rule

1 yd = 3 ft

yd	ft
2	
	9
5	
	30

6. Solve.

Unit

22 − 14 = _____

62 − 14 = _____

162 − 14 = _____

_____ = 292 − 14

_____ = 402 − 14

Date Time

Things to Measure

1. List 5 things you can measure with a ruler, a tape measure, a meterstick, or a yardstick.

____

____

____

____

____

2. List 5 things you can weigh with a scale.

____

____

____

____

____

3. List 5 things you can measure with a measuring cup, a measuring spoon, or some other container.

____

____

____

____

____

Date Time

Math Boxes 9.6

1. Write the number that is 100 less.

465 ______

700 ______

960 ______

4,391 ______

2. Write 5 names for $1.50.

______

______

______

______

______

3. Circle the unit that makes sense.

Grandma's house is

5 _____ away. km dm

Amy's goldfish is

8 _____ long. cm m

Ahmed's dad is

68 _____ tall. in. cm

4. Solve.

Unit
km

6 + 5 = ______

60 + 50 = ______

600 + 500 = ______

6,000 + 5,000 = ______

5. Measure each side to the nearest cm. Find the perimeter.

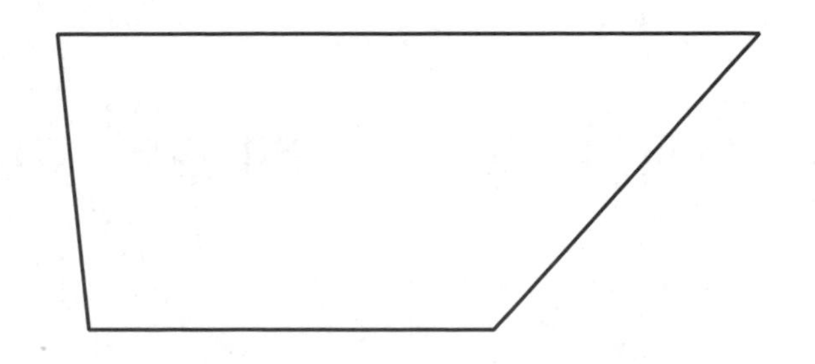

The perimeter is ______ cm.

6. Trade first. Then solve.

Unit
miles

$$\begin{array}{r} 87 \\ -\ 39 \\ \hline \end{array}$$

Date Time

Math Message

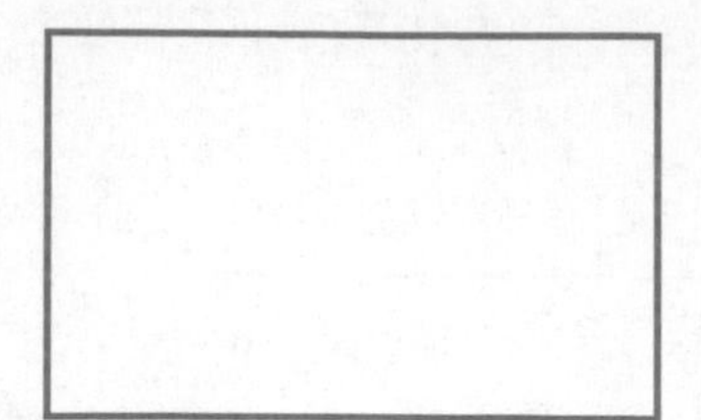

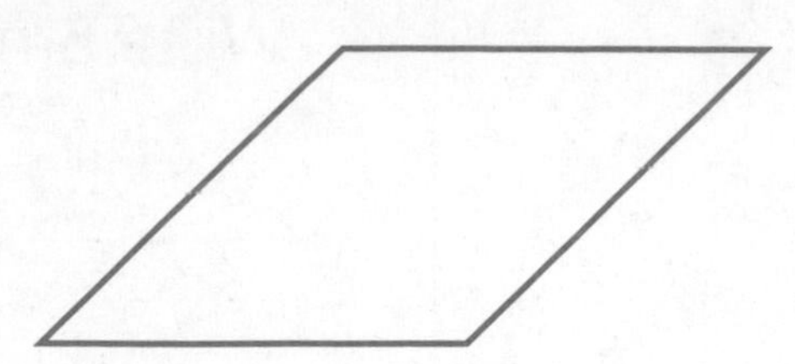

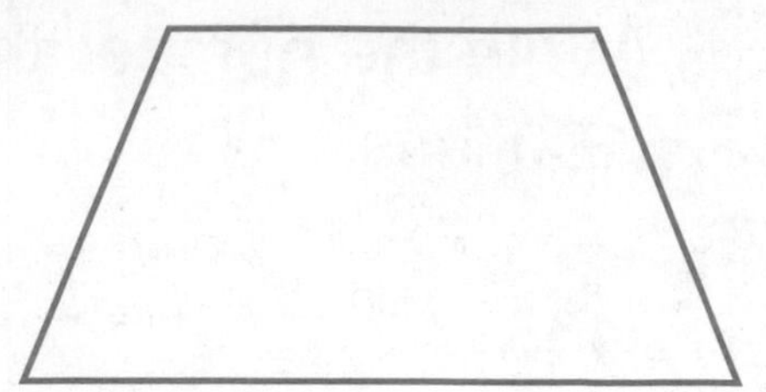

Estimate: Which shape is the "biggest" (has the largest area)? Circle it.

Think: How might you measure the shapes to find out?

Exploration A: Which Cylinder Holds More?

Which holds more macaroni—the tall and narrow cylinder or the short and wide cylinder?

My prediction: ______________________________

Actual result: ______________________________

Exploration B: Measuring Area

The area of my tracing of the deck of cards is about _____ square centimeters.

The area of my tracing of the deck of cards is about _____ square inches.

I traced ______________________________.

It has an area of about ______________________________.
(unit)

Math Boxes 9.7

1. Complete the puzzle.

Start

175

End

2. Use counters to find the answer.

I grabbed $\frac{1}{2}$ of the pile of pennies. I have 6.

How many pennies in all?

_____ pennies

How many pennies are left?

_____ pennies

3. Use counters to solve.

18 orange slices are shared equally. Each child gets 4 slices.

How many children are sharing?

_____ children

How many slices are left?

_____ slices

4. The time is 10:20. Draw hands to show $\frac{1}{2}$ hour later.

The time will be

_____:_____.

5. Sherry jumped 29 inches. Her sister doubled her jump. How far did her sister jump?

_____ inches

6. Color $\frac{2}{3}$ of the leaves green.

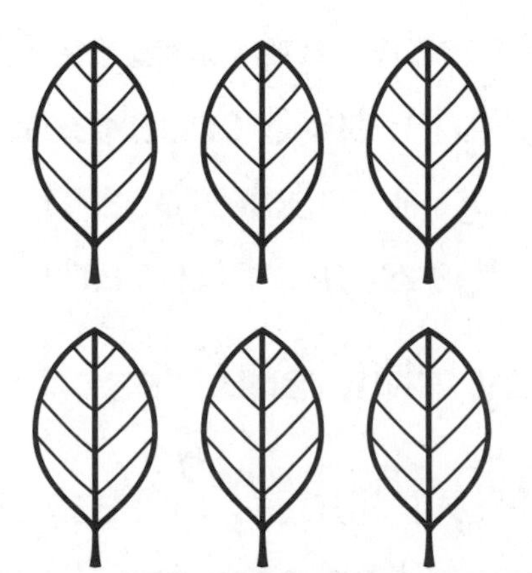

Date Time

Letter Areas

Letter Areas Find the area of each letter.

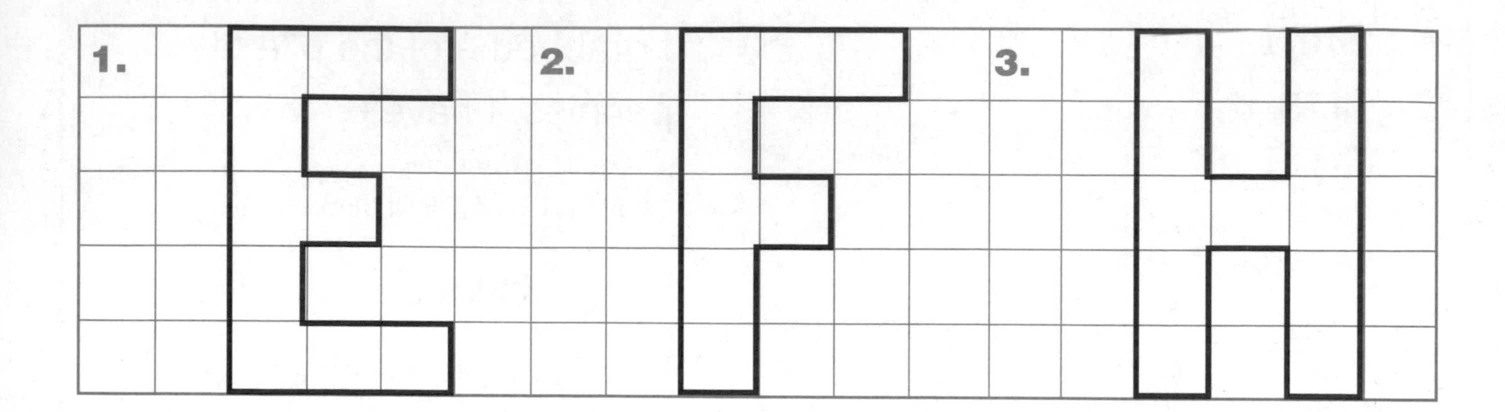

Area = ____ sq cm **Area** = ____ sq cm **Area** = ____ sq cm

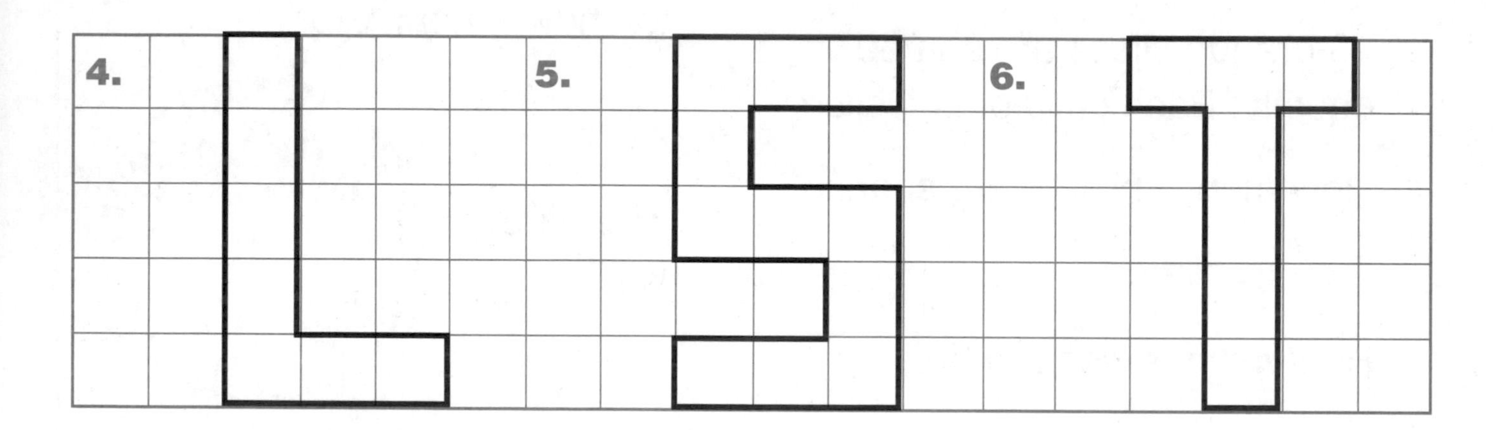

Area = ____ sq cm **Area** = ____ sq cm **Area** = ____ sq cm

Challenge An 8-by-8 checkerboard has 64 squares. Some squares on a checkerboard are white. Some are black. Squares of the same color are never next to each other.

7. How many white squares are there in each row? _____
   How many black squares? _____

8. If the square in one corner is black, what color is the square in the opposite corner? _______

Date Time

Math Boxes 9.8

1. Write 4 names for $\frac{1}{2}$. Use your Fraction Cards to help.

_____, _____, _____, _____

Write 2 fractions that are:

greater than $\frac{1}{2}$. _____, _____

less than $\frac{1}{2}$. _____, _____

2. Solve.

Unit

$8 + 6 =$ _____

$8 + 6 + 7 =$ _____

$8 + 6 + 7 + 5 =$ _____

$8 + 6 + 7 + 5 + 9 =$ _____

3. Count by quarters to $3.00.

$0.50, _____, _____,

_____, _____, _____,

_____, _____, _____,

_____, _____

4. Draw two ways to show $\frac{2}{3}$.

5. Circle $\frac{3}{8}$.

6. 5 nickels = 1 quarter

1 nickel = $\frac{1}{5}$ of a quarter

_____ inches = 1 yard

1 inch = _____ of a yard

_____ inches = 2 feet

1 inch = _____ of 2 feet

Date Time

Equivalent Units of Capacity

Complete.

U.S. Customary Units of Capacity

______ pint = 1 cup

1 pint = ______ cups

______ pints = 1 quart

______ quarts = 1 half-gallon

______ half-gallons = 1 gallon

Metric Units of Capacity

1 liter = ______ milliliters

$\frac{1}{2}$ liter = ______ milliliters

1. How many quarts are in 1 gallon? ______ quarts
2. How many cups are in 1 quart? ______ cups

   In a half-gallon? ______ cups   In 1 gallon? ______ cups
3. How many pints are in a half-gallon? ______ pints

   In a gallon? ______ pints

"What's My Rule?"

4. **Rule**

   1 qt = 2 pt

qt	pt
2	
3	
	10
8	

5. **Rule**

   1 gal = 8 pt

gal	pt
2	
3	
	40
	80

Date Time

Math Boxes 9.9

1. Solve.

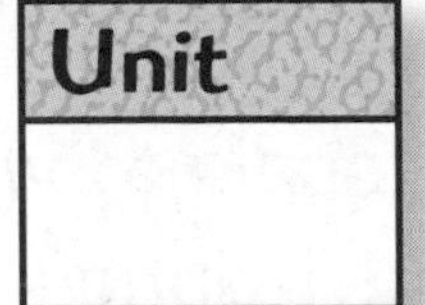

$$\begin{array}{r} 17 \\ +\ 8 \\ \hline \end{array} \qquad \begin{array}{r} 17 \\ 8 \\ +\ 5 \\ \hline \end{array} \qquad \begin{array}{r} 17 \\ 8 \\ 5 \\ +\ 19 \\ \hline \end{array}$$

2. Write $<$, $>$, or $=$.

$4 + 5 + 6$ _____ $3 + 5 + 7$

$7 + 5 + 9$ _____ $6 + 6 + 8$

$2 + 11 + 4$ _____ $7 + 1 + 9$

$15 + 7 + 5$ _____ $9 + 9 + 9$

3. Draw a square with a perimeter of 8 cm.

Remember: The sides of a square are equal.

4. Draw two ways to show $\frac{3}{4}$.

5. Make a square array with 25 pennies. How many pennies in each row?

_____ pennies

6. Complete the diagram. Then write a number model.

Start	Change	End
54	+38	

____

Date Time

Weight

Weighing Pennies

Use a spring scale, letter scale, or diet scale to weigh pennies. Find the number of pennies that weigh about 1 ounce.

I found that ______ pennies weigh about 1 ounce.

Which Objects Weigh about the Same?

Work in a small group. Your group will be given several objects that weigh less than 1 pound.

1. Choose two objects. Hold one object in each hand and compare their weights. Try to find two objects that weigh about the same.

   Two objects that weigh about the same:

   ______________________ ______________________

2. After everyone in the group has chosen two objects that weigh about the same, weigh all of the objects. Record the weights below.

Object	**Weight** (include the unit)
______________________	______________
______________________	______________
______________________	______________
______________________	______________

Date Time

Thinking about Weight

For Problems 1–4, fill in the blanks.

1. Are all boxes of cereal that weigh the same also the same size and shape?

   ____

2. Which weighs more, 1 pound of potato chips or 1 pound of potatoes?

   ____

   Which takes up more space?

   ____

3. A phone book for the city of Chicago weighs about 4 pounds.

   I weigh about ________ pounds.

   I weigh about ________ times as much as a Chicago phone book.

4. If 1 pound of bananas costs 50 cents, how much do 2 pounds cost?

   ____

   How much does half of a pound cost? ________

   How much do $1\frac{1}{2}$ pounds cost? ________

Date Time

Thinking about Weight (cont.)

Circle the best answer.

5. A small dog might weigh about …

4 feet 4 pounds 4 gallons 4 grams

6. A letter might weigh about …

5 ounces 5 grams 5 meters 5 kilograms

7. A bunch of 5 bananas might weigh about …

3 ounces 3 grams 3 liters 3 pounds

8. A whole watermelon might weigh about …

6 kilograms 6 ounces 6 quarts 6 decimeters

Date Time

Math Boxes 9.10

1. The total cost is 60¢. You pay with a $1 bill.

   How much change do you get?

   _____

   Show the change using Ⓠ, Ⓓ, and Ⓝ.

2. Show 8 groups of 2 □s.

   How many □s in all? _____

3. Estimate. Then solve.

   Jen drove 127 miles the first day. She drove 154 miles the second day. How many miles did she drive in all?

   Estimate: _____ miles

   Answer: _____ miles

4. (grid with rectangle)

   Area = _____ sq cm

   Perimeter = _____ cm

5. **Rule**

   10 cm = 1 dm

cm	dm
5	
30	
	6

6. Match.

5 ft	3 yd
24 in.	60 in.
9 ft	2 ft

Date Time

Math Boxes 9.11

1. Solve.

Unit

$9 + 7 =$ _____

$9 + 7 + 8 =$ _____

$9 + 7 + 8 + 5 =$ _____

$9 + 7 + 8 + 5 + 6 =$ _____

2. Write 2 even 4-digit numbers.

_________ _________

Write 2 odd 4-digit numbers.

_________ _________

3. Solve.

Unit
dinosaurs

_____ $= 7 + 9$

_____ $= 37 + 9$

$16 - 8 =$ _____

$76 - 8 =$ _____

4. Label each part with a fraction.

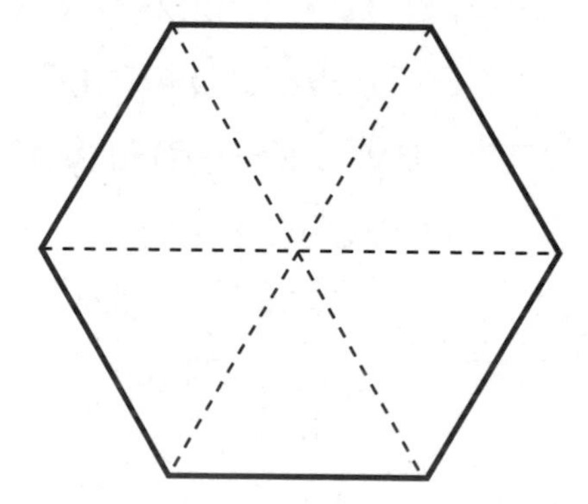

5. How much in all?

____

6. Circle the unit that makes sense.

A can of soup may weigh:

8 ounces 8 cups

8 pounds 8 feet

A loaf of bread may weigh:

1 pound 1 quart

1 gram 8 liters

Date Time

Math Boxes 10.1

1. Fill in the missing numbers.

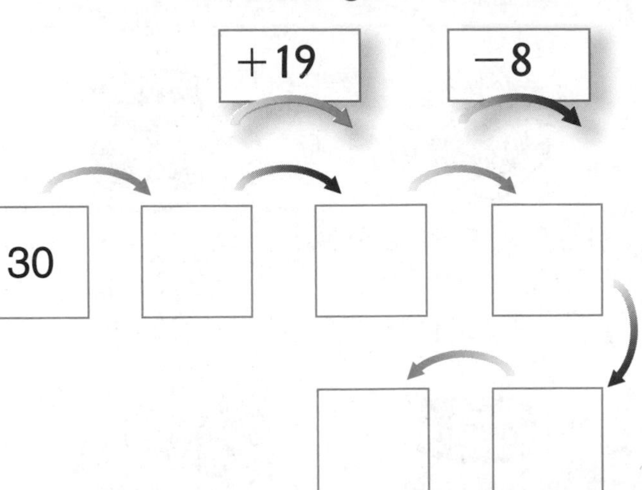

2. 264 246 277

301 310

Unit
meters

The median number of meters is _____.

3. Use <, >, or =.

Unit
lunches

37 _____ 5 + 5 + 20

20 + 7 + 50 _____ 77

14 _____ 12 + 4 + 6

52 _____ 25 + 25 + 4

4. Make a square array with 36 pennies. How many pennies in each row?

_____ pennies

5.

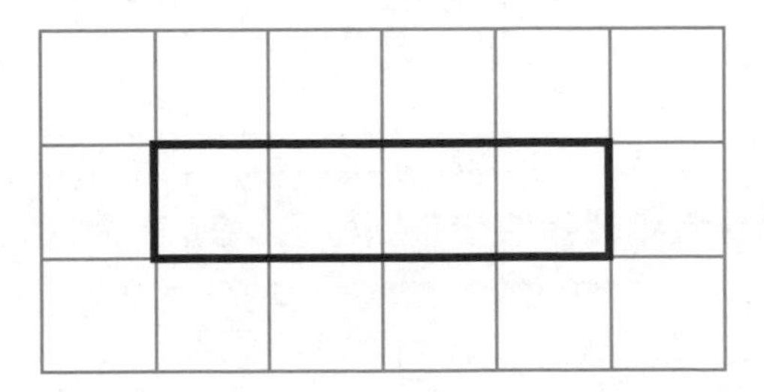

The area is _____ sq cm.

The perimeter is _____ cm.

6. Solve.

Unit

70 + 80 = _____

_____ = 150 − 90

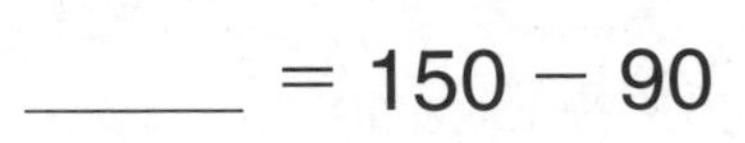

90 + 80 = _____

160 − 70 = _____

Date Time

Good Buys Poster

Fruit/Vegetables Group

Seedless Grapes
99¢ lb

Carrots
1-lb bag
3/$1.00

Plums
69¢ lb

Oranges
$1.49 lb

Bananas
59¢ lb

Watermelons
$2.99 ea.

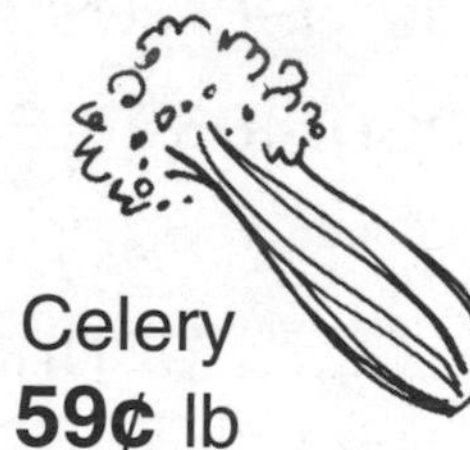

Celery
59¢ lb

Meat Group

U.S.D.A. Choice Fresh
Ground Beef
$1.99 lb

Peanut Butter
18-oz jar
$1.29

Chunk Light
Tuna
6.5 oz
69¢

Lunch Meat
1-lb package
$1.39

Milk Group

Gallon
Milk
$2.39

American
Cheese
8 oz
$1.49

6-pack
Yogurt
$2.09

Grain Group

Wheat Bread
16 oz
99¢

Potato Chips
8 oz
89¢

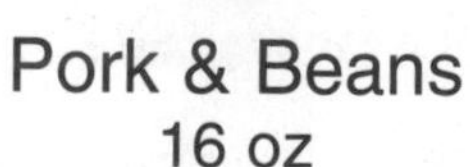

Pork & Beans
16 oz
2/89¢

Saltines
1 lb
69¢

Hamburger Buns
16 oz
69¢

Miscellaneous Items

Mayonnaise
32 oz
$1.99

Catsup
32 oz
$1.09

Grape Jelly
2-lb jar
$1.69

Use with Lesson 10.1.

Date Time

Ways to Pay

Choose 4 items from the Good Buys Poster on page 240. List the items and how much they cost in the table below.

For each item:

- Count out coins and bills to show several different ways of paying for each item.
- Record two ways by drawing coins and bills in the table. Use Ⓠ, Ⓓ, Ⓝ, Ⓟ, and [$1].

Example

You buy 1 pound of bananas. They cost 59¢ a pound. You pay with:

Ⓠ Ⓠ Ⓝ Ⓟ Ⓟ Ⓟ Ⓟ or Ⓓ Ⓓ Ⓓ Ⓓ Ⓓ Ⓝ Ⓟ Ⓟ Ⓟ Ⓟ

1. You buy ______________. Cost: ______________ Pay with	or	**2.** You buy ______________. Cost: ______________ Pay with	or
3. You buy ______________. Cost ______________ Pay with	or	**4.** You buy ______________. Cost: ______________ Pay with	or

Date Time

Word Values

Pretend the letters of the alphabet have the dollar values shown in the table. For example, the letter **g** is worth $7; the letter **v** is worth $22. The word **jet** is worth $10 + $5 + $20 = $35.

	a	b	c	d	e	f	g	h	i	j	k	l	m
Value	$1	$2	$3	$4	$5	$6	$7	$8	$9	$10	$11	$12	$13
	n	**o**	**p**	**q**	**r**	**s**	**t**	**u**	**v**	**w**	**x**	**y**	**z**
Value	$14	$15	$16	$17	$18	$19	$20	$21	$22	$23	$24	$25	$26

1. Which is worth more, **dog** or **cat**? ______
2. Which is worth more, **whale** or **zebra**? ______
3. How much is your first name worth? ______
4. Write 2 spelling words you are trying to learn. Find their values.

   Word: ______ Value: $______

   Word: ______ Value: $______
5. What is the cheapest word you can make? It must have at least 2 letters.

   Word: ______ Value: $______
6. What is the most expensive word you can make?

   Word: ______ Value: $______
7. Think of the letter values as dimes. For example, **m** is worth 13 dimes; **b** is worth 2 dimes. Find out how much each word is worth.

   dog: $______ cat: $______ zebra: $______ whale: $______

   candy: $______ your last name: $______

Date Time

Math Boxes 10.2

1. I have 75¢. How many 20¢ erasers can I buy?

   _____ erasers

2. I have a pile of pennies. $\frac{1}{3}$ of the pile is 6 pennies. How many pennies are in the whole pile?

   _____ pennies

3. Circle the fraction that is more. Use your Fraction Cards to help.

   $\frac{2}{3}$ or $\frac{2}{2}$

   $\frac{4}{5}$ or $\frac{2}{5}$

   $\frac{2}{8}$ or $\frac{5}{6}$

   $\frac{3}{6}$ or $\frac{1}{4}$

4. Draw a triangle. Measure each side to the nearest inch.

   about _____ in.

   about _____ in.

   about _____ in.

5. It is 6:15. Draw the hour and minute hands to show the time 15 minutes later.

   What time does the clock show?

   _____ : _____

6. _____ pennies = \$2.00

   _____ nickels = \$2.00

   _____ dimes = \$2.00

   _____ quarters = \$2.00

Date Time

Calculator Dollars and Cents

To enter $4.27 into your calculator, press (4) (.) (2) (7).

To enter 35¢ into your calculator, press (.) (3) (5).

1. Enter $3.58 into your calculator. The display shows ____________.

2. Enter the following amounts into your calculator.
Record what the display shows.
Don't forget to clear between each entry.

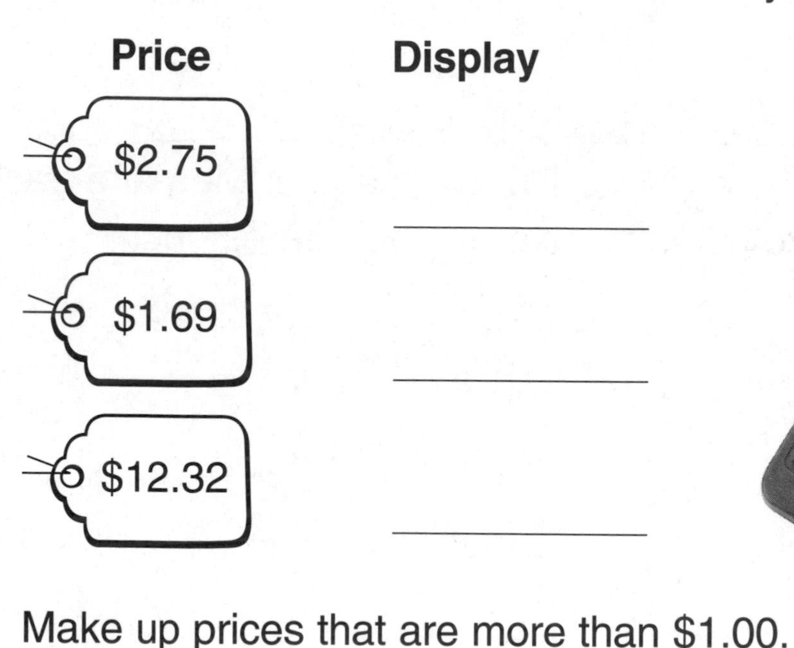

Price	Display
$2.75	____________
$1.69	____________
$12.32	____________

Make up prices that are more than $1.00.

Price	Display
	____________
	____________
	____________

3. Enter 68¢ into your calculator. The display shows ____________.

Date Time

Calculator Dollars and Cents (cont.)

4. Enter the following amounts into your calculator. Record what you see in the display.

Price	Display
$0.10	____________
$0.26	____________
$0.09	____________

Make up prices that are less than $1.00.

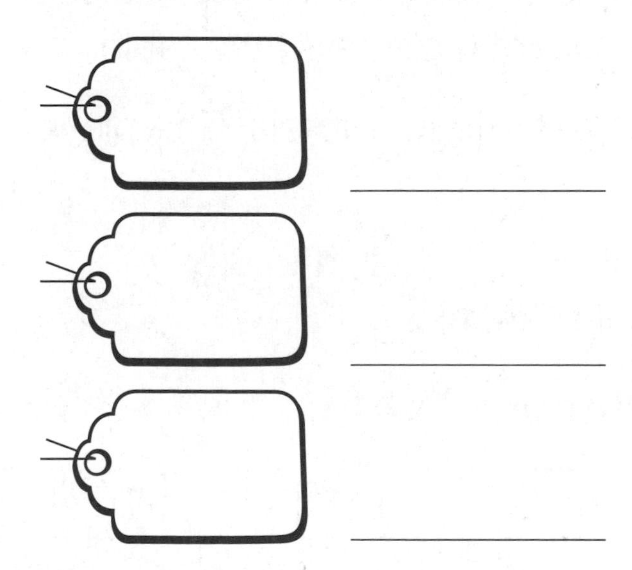

5. Use your calculator to add $1.55 and $0.25.

What does the display show? ______

Explain what happened. ____

____

____

Date Time

Pick-a-Coin

Materials ❑ 1 die ❑ calculator for each player

❑ *Pick-a-Coin* record table for each player (*Math Journal 2,* p. 247)

Players 2 to 4

Summary

Players roll a die. The numbers that come up are used as numbers of coins and dollar bills. Players try to make collections of coins and bills with the largest value.

Directions

Take turns. When it is your turn, roll the die five times. After each roll, record the number that comes up on the die in any one of the empty cells in the row for that turn on your record table. Then use a calculator to find the total amount for that turn. Record the total in the table.

After four turns, use your calculator to add the four totals. The player with the largest Grand Total wins.

Example: On his first turn, Brian rolled 4, 2, 4, 1, and 6. He filled in his record table like this:

Pick-a-Coin Record Table

	Ⓟ	Ⓝ	Ⓓ	Ⓠ	$1	Total
1st turn	2	1	4	4	6	$7.47
2nd turn						$___.____
3rd turn						$___.____
4th turn						$___.____
				Grand Total		$___.____

Date Time

Pick-a-Coin Record Tables

	Ⓟ	Ⓝ	Ⓓ	Ⓠ	$1	Total
1st turn						$___ . _____
2nd turn						$___ . _____
3rd turn						$___ . _____
4th turn						$___ . _____
					Grand Total	$___ . _____

	Ⓟ	Ⓝ	Ⓓ	Ⓠ	$1	Total
1st turn						$___ . _____
2nd turn						$___ . _____
3rd turn						$___ . _____
4th turn						$___ . _____
					Grand Total	$___ . _____

	Ⓟ	Ⓝ	Ⓓ	Ⓠ	$1	Total
1st turn						$___ . _____
2nd turn						$___ . _____
3rd turn						$___ . _____
4th turn						$___ . _____
					Grand Total	$___ . _____

Date Time

Math Boxes 10.3

1. If \$1.00 is ONE, then

   1¢ = 1/100

   25¢ = ____

   32¢ = ____

   50¢ = ____

   99¢ = ____

2. Draw 12 Ⓠs. Circle $\frac{1}{3}$ of them. What is the value of the circled coins?

   \$______

3. Rosita had \$0.39 and found \$0.57 more. How much does she have now? Estimate your answer and then use partial sums to solve.

   Estimate:

   ____ + ____ = ____

   Answer: ________

4. Solve.

Unit

   50 + 30 = ____

   40 + 80 = ____

   ____ = 70 + 70

   ____ = 20 + 90

5. Use Ⓟ, Ⓝ, Ⓓ, and Ⓠ. Show \$1.79 two different ways.

6. 17 magazines are shared equally among 5 children. Draw a picture to help you.

   Each child gets ____ magazines.

   There are ____ magazines left.

Math Boxes 10.4

1. Write $<$, $>$, or $=$.

$0.73 _____ $0.07

$0.46 _____ $1.46

$3.29 _____ $2.93

2. Draw 16 Ⓠs. Circle $\frac{1}{4}$ of them. What is the value of the circled coins?

$________

3. Write the money amounts in dollars-and-cents notation.

= $______ = $______

= $______ = $______

= $______

4. I have $2.00. Can I buy 4 bags of chips for $0.55 each?

______

5. Fill in the diagram and write a number model.

Start	Change	End
27	+27	

____

6. Estimate the height of your desk. Then measure it.

Estimate: about ______________
(unit)

Measurement: ______________
(unit)

Date Time

Then-and-Now Poster

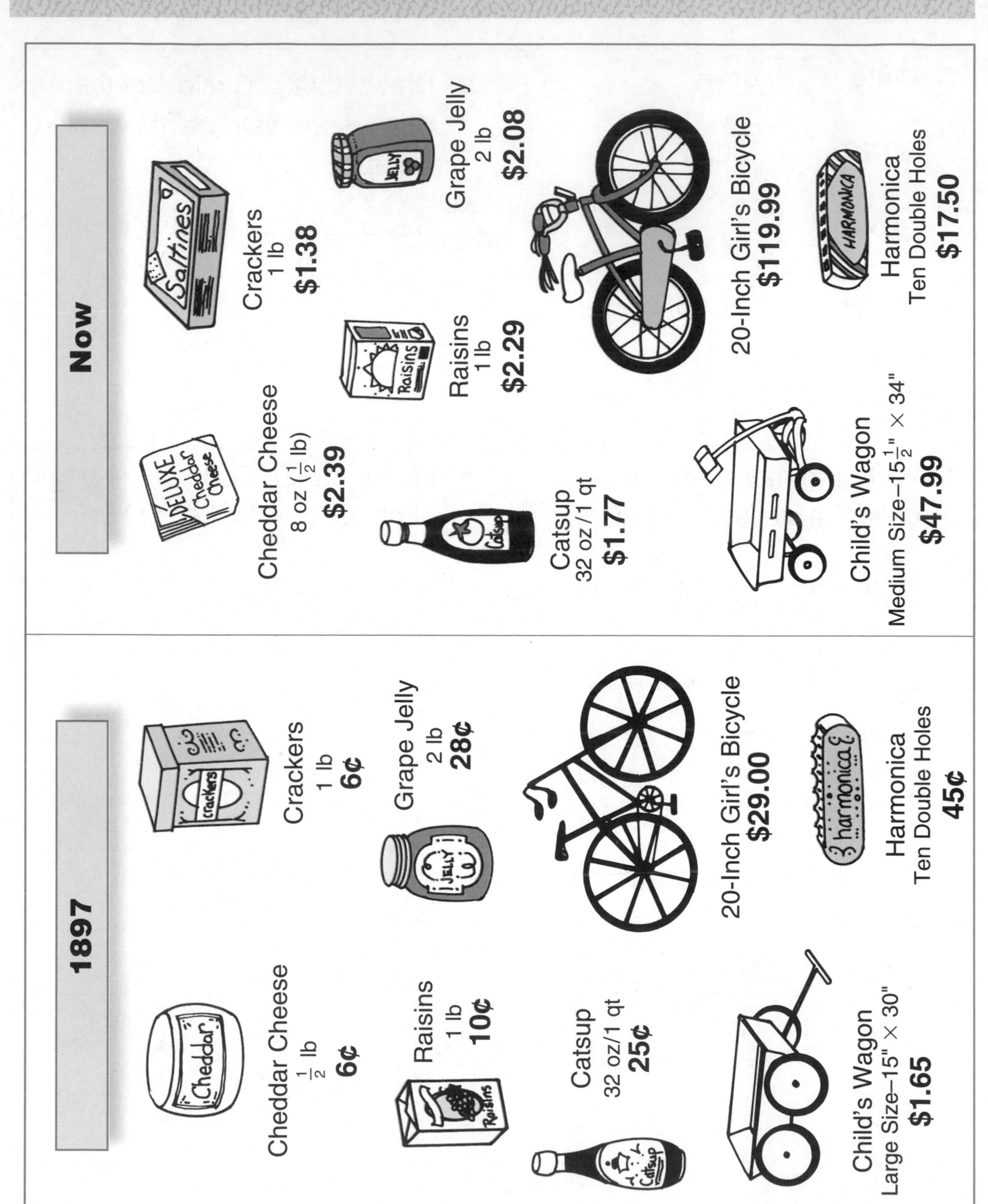

Date Time

Then-and-Now Prices

Use your calculator.

1. How much did a 20-inch bicycle cost in 1897? ___________

   How much does it cost now? ___________

   How much more does it cost now? ___________

2. How much more does a pound of cheese cost now than it did in 1897? ___________

3. In 1897, raisins were packed in cartons. Each carton contained 24 one-pound boxes. How much did a 24-pound carton cost then? ___________

   How much would it cost now? ___________

4. Which item had the biggest price increase from "then" to "now"?

   ________________________ had the biggest price increase.

   How much more does it cost now? ___________

5. Our Own Problems about Then-and-Now:

   ____

   ____

   ____

   ____

   ____

   ____

Date Time

Buying Food

Choose items to buy from the Good Buys Poster on journal page 240.
For each purchase:

- Record the items on the sales slip.
- Write the price of each item on the sales slip.
- Estimate the total cost and record it.
- Find the total cost and write it on the sales slip.

Purchase 1 **Store**

Items:

________________ $___ . _____

________________ $___ . _____ Estimated cost:

Total: $___ . _____ about $___ . _____

Purchase 2 **Store**

Items:

________________ $___ . _____

________________ $___ . _____ Estimated cost:

Total: $___ . _____ about $___ . _____

Purchase 3 **Store**

Items:

________________ $___ . _____

________________ $___ . _____ Estimated cost:

Total: $___ . _____ about $___ . _____

Date Time

Math Boxes 10.5

1. Estimate. Then use partial sums to solve.

Estimate:

_____ + _____ = _____

$$\begin{array}{r} 57 \\ +\ 48 \\ \hline \end{array}$$

2. Solve.

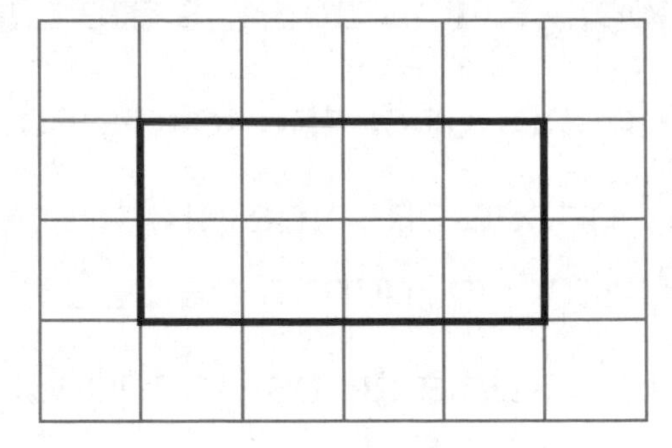

Area: _____ sq cm

Perimeter: _____ cm

3. There are _____ pennies in 1 dime. What fraction of a dime is 1 penny? _____

There are _____ pennies in 1 quarter. What fraction of a quarter is 1 penny? _____

4. The pet store sold 12 fish. $\frac{1}{2}$ were guppies and $\frac{1}{4}$ were neons. The rest were angelfish. How many of each?

There were _____ guppies.

There were _____ neons.

There were _____ angelfish.

5. _____ pennies = \$3.00

_____ nickels = \$3.00

_____ dimes = \$3.00

_____ quarters = \$3.00

6. Circle the answer.

\$2.88 is closer to:

\$2.80 or \$2.90

\$5.61 is closer to:

\$5.60 or \$5.70

\$1.97 is closer to:

\$1.90 or \$2.00

Making Change

Work with a partner. Use your tool-kit coins and bills. One of you is the Shopper. The other is the Clerk.

The Shopper does the following:

- Chooses one item from each food group on the Good Buys Poster on journal page 240.
- Lists these items on the Good Buys sales slip in the Shopper's journal on page 255.
- Writes the cost of each item on the sales slip.
- Estimates the total cost of all the items and writes it on the sales slip.
- Pays with a $10 bill.
- Estimates the change and writes it on the sales slip.

The Clerk does the following:

- Uses a calculator to find the exact total cost.
- Writes the exact total cost on the sales slip.
- Gives the Shopper change by counting up.
- Writes the exact change from $10.00 on the sales slip.

The Shopper uses a calculator to check the Clerk's change.

Change roles and repeat.

Date Time

Making Change (cont.)

The Good Buys Store Sales Slip		
	Item	**Cost**
Fruit/vegetables group	________	$ ___ . ____
Grain group	________	$ ___ . ____
Meat group	________	$ ___ . ____
Milk group	________	$ ___ . ____
Miscellaneous items	________	$ ___ . ____
Estimated total cost		$ ___ . ____
Estimated change from $10.00		$ ___ . ____
Exact total cost		$ ___ . ____
Exact change from $10.00		$ ___ . ____

Date Time

Math Boxes 10.6

1. Cross out the names that don't belong.

10¢

ten cents, $\frac{1}{10}$ of a dollar,

\$10.00, Ⓓ, ⓃⓃ, \$0.01,

$\frac{1}{100}$ of a dollar,

$\frac{1}{2}$ of a dollar

2. A $\frac{1}{2}$ pint of berries costs \$0.99. I have \$2.00. Can I buy 1 pint of berries?

______

3. Solve.

Unit

180 − 60 = ______

150 + 60 + 40 = ______

240 − 120 = ______

170 + 30 + 80 = ______

4. Trade first. Then subtract.

$$\begin{array}{r} \$0.81 \\ -\ \$0.35 \\ \hline \end{array}$$

5. If \$1.00 is ONE, then

10¢ = $\frac{1}{10}$

30¢ = ______

50¢ = ______

80¢ = ______

\$2.00 = ______

6. Write >, <, or =.

1 qt ______ 1 pt

3 c ______ 1 gal

1 qt ______ 4 c

1 gal ______ 5 pt

Date Time

The Area of My Handprint

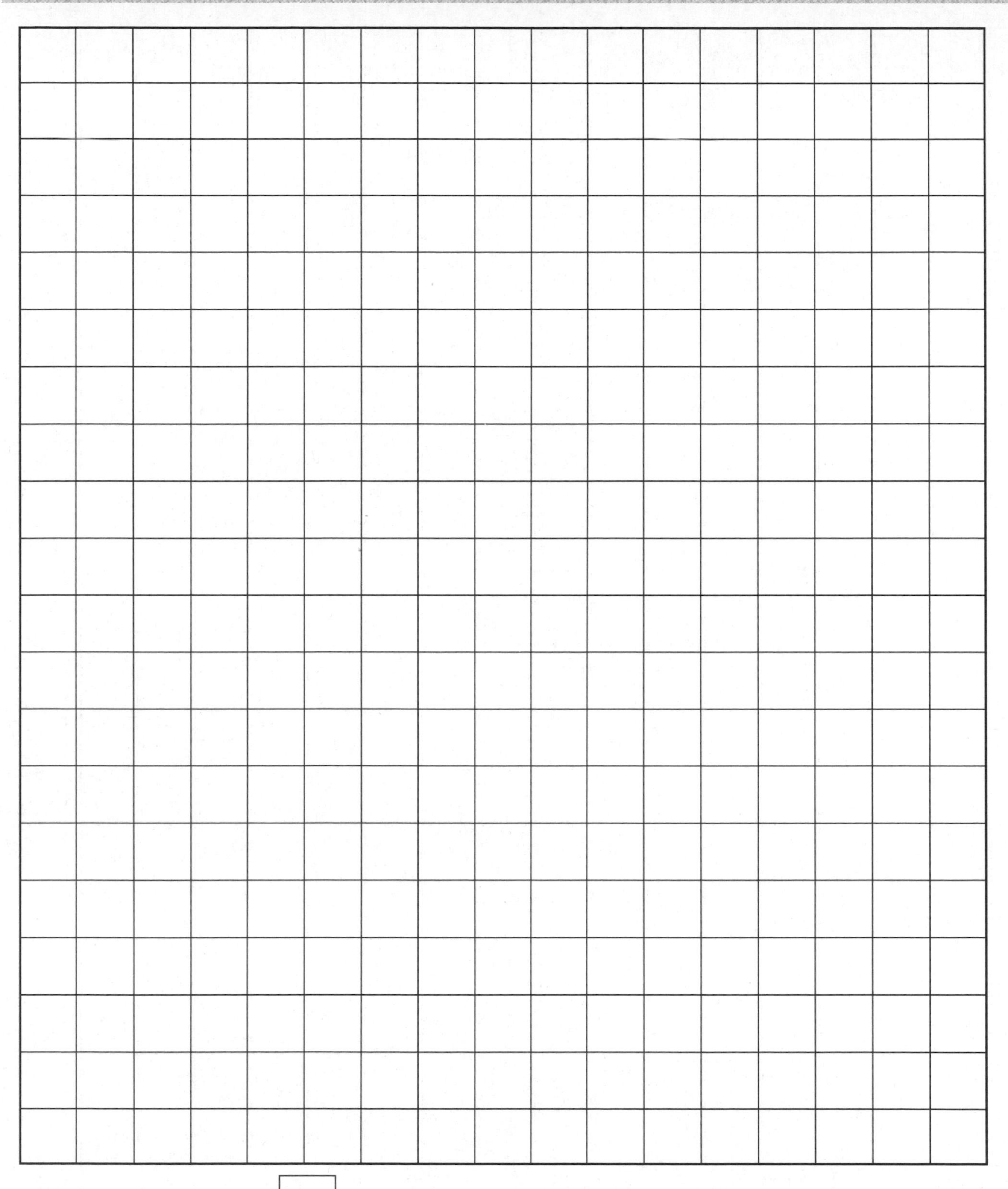

The area of each ☐ is 1 square centimeter. Other ways to write *square centimete*r are sq cm and cm^2.

The area of my handprint is _____ square centimeters, or _____ sq cm.

Date Time

The Area of My Footprint

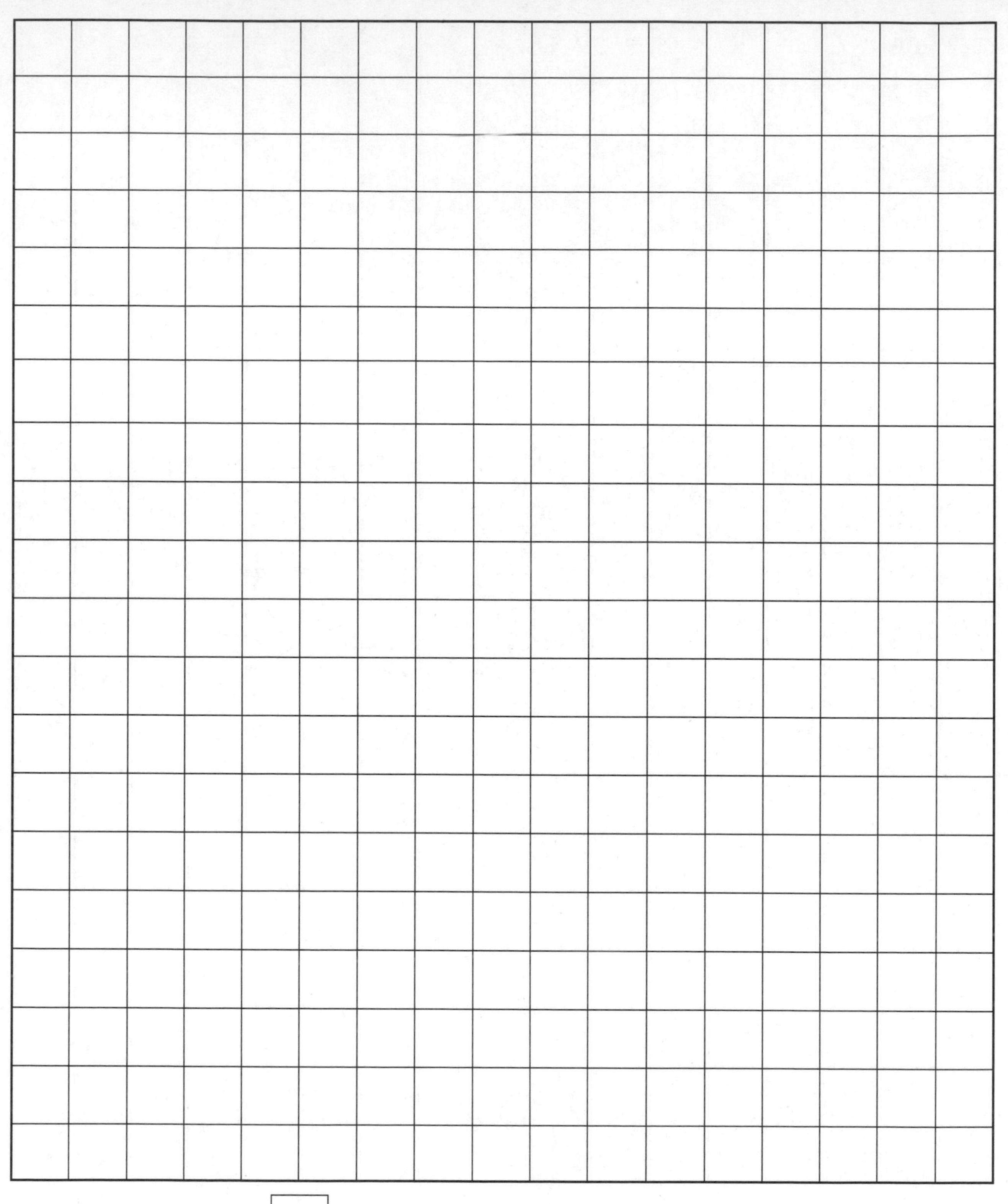

The area of each ☐ is 1 square centimeter. Other ways to write *square centimeter* are sq cm and cm^2.

The area of my footprint is _____ square centimeters, or _____ sq cm.

Date Time

Worktables

Use a Pattern-Block Template to draw each of the different-size and different-shape worktables you made. Write the name of each shape next to your drawing.

Date Time

Geoboard Dot Paper

1.

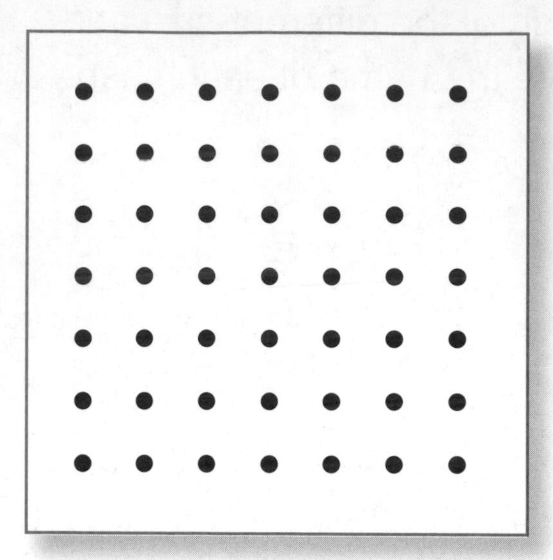

2.

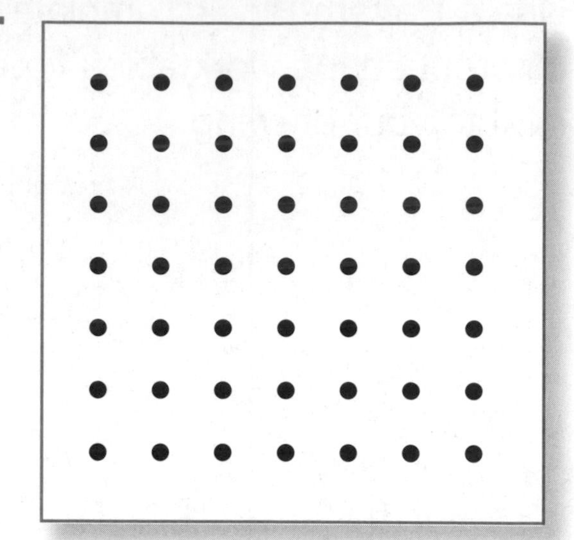

3.

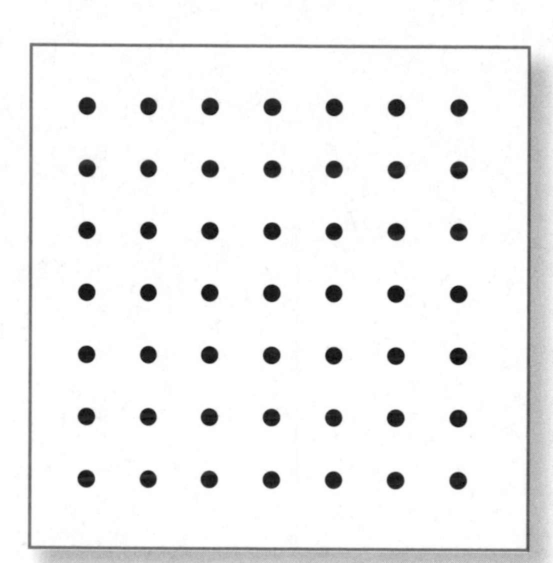

4.

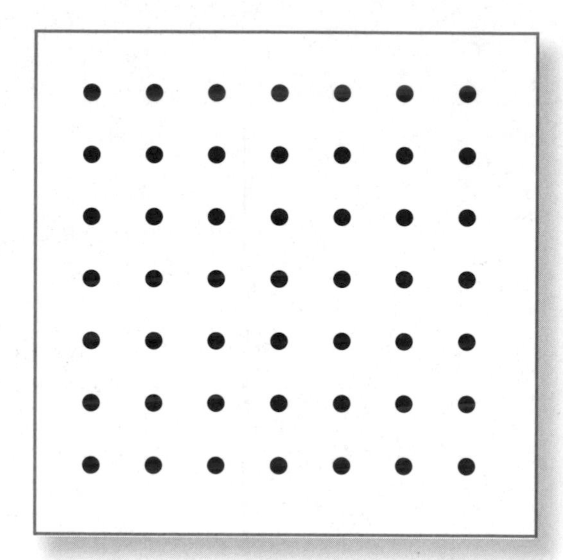

5.

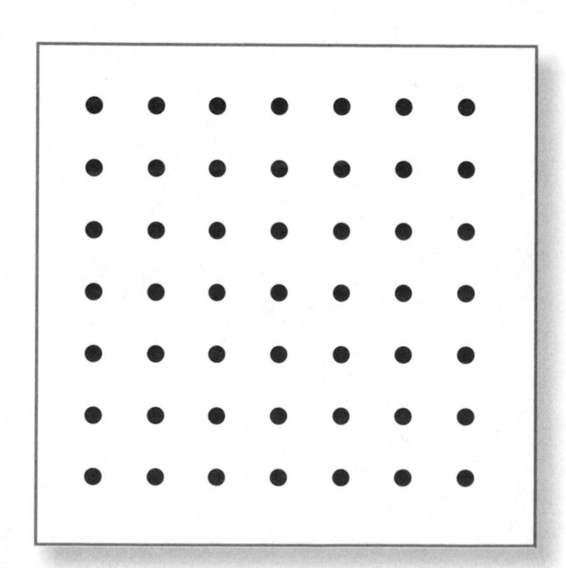

6.

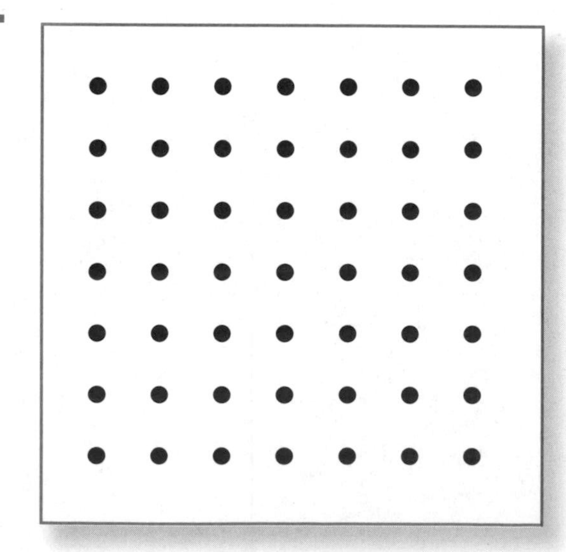

Use with Lesson 10.7.

Date Time

Math Boxes 10.7

1. Fill in the blanks to estimate the total cost.

   $2.43 + $0.39 is about

   ______ + ______ = ______

   $0.88 + $0.67 is about

   ______ + ______ = ______

2. I have a 5-dollar bill. I spend $4.38. How much change do I get?

   ______

3. Write as dollars and cents.

   eight dollars and forty-three cents: ______

   fifteen dollars and 6 cents:

   ______

   fifty dollars and seventeen cents: ______

4. Use a straightedge. Draw a rectangle. Measure the sides to the nearest inch.

   about ______ in.

   about ______ in.

   about ______ in.

   about ______ in.

5. A cube is the ONE. What number is shown by the blocks?

   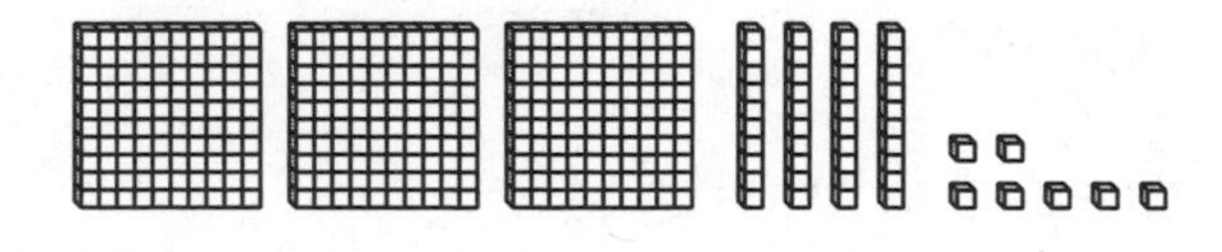

   ______

6. Color $\frac{3}{4}$ of the circle.

   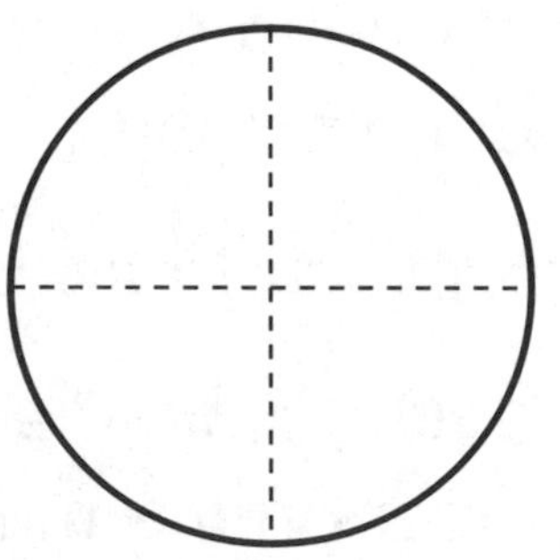

   What fraction of the circle is not colored? ______

Date Time

Money Exchange Game

Materials
- ❑ 1 six-sided die
- ❑ 1 ten- or twelve-sided die
- ❑ 24 pennies, 39 dimes, thirty-nine $1 bills, and one $10 bill per player

Players 2 or 3

Directions

1. Each player puts 12 pennies, 12 dimes, twelve $1 bills, and one $10 bill in the bank.
2. Players take turns. Players use a six-sided die to represent pennies. Players use a ten- or twelve-sided die to represent dimes.
3. Each player
   - rolls the dice.
   - takes from the bank the number of pennies and dimes shown on the faces of the dice.
   - puts the coins in the correct columns on his or her Place-Value Mat on journal page 263.
4. Whenever possible, a player replaces 10 coins or bills of a lower denomination with a coin or bill of the next higher denomination.
5. The first player to trade for a $10 bill wins.

If there is a time limit, the winner is the player with the largest number on the mat when time is up.

Date Time

Place-Value Mat

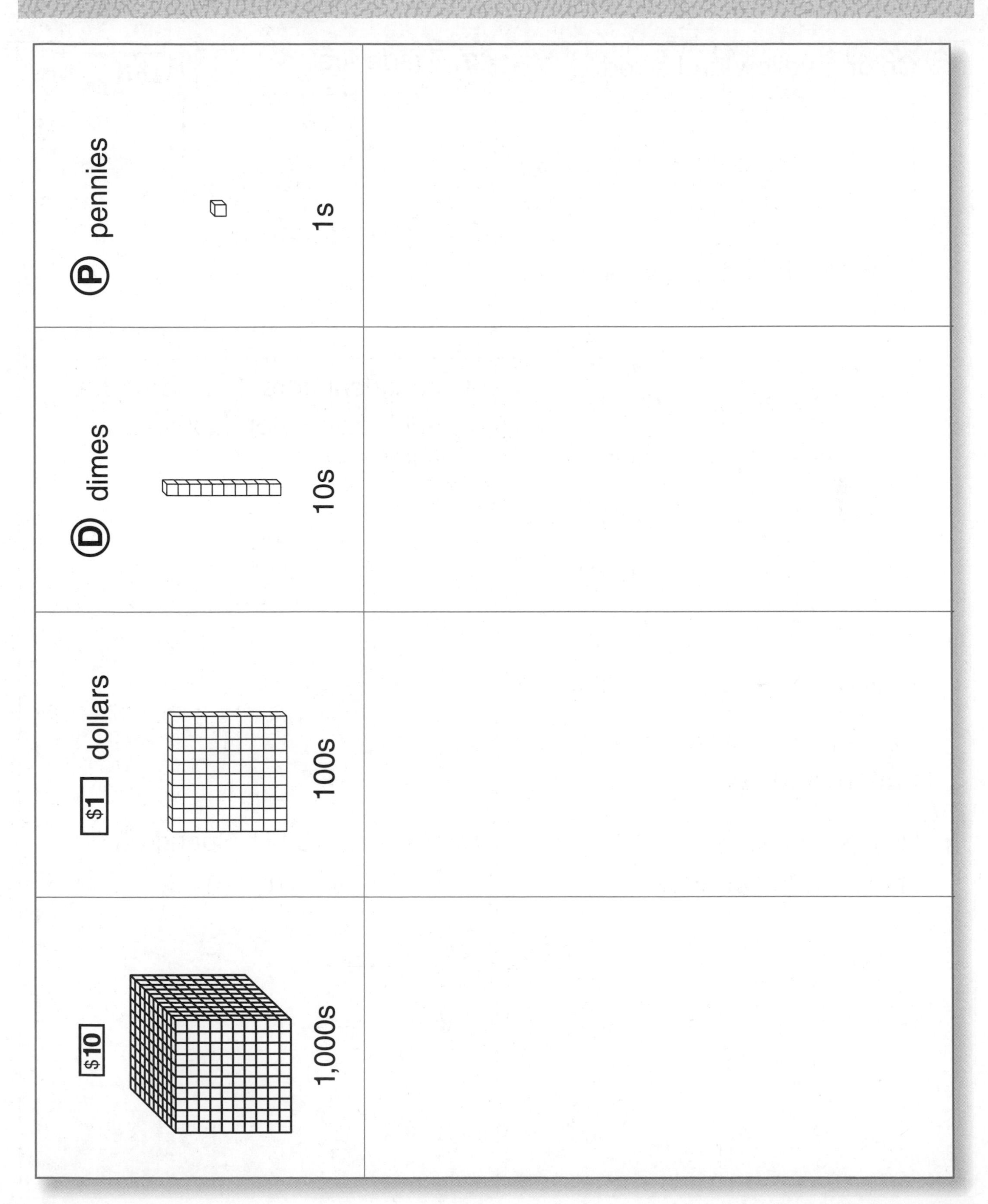

Date Time

Math Boxes 10.8

1. Color $\frac{2}{8}$ yellow and $\frac{5}{8}$ red.

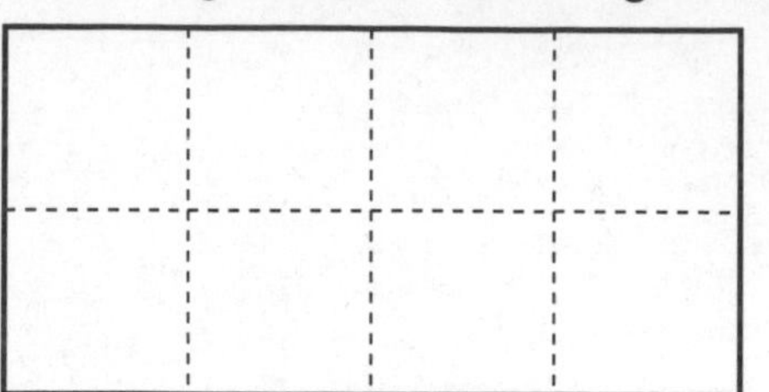

How much is not colored?

______

2. Trade first.
Then subtract.

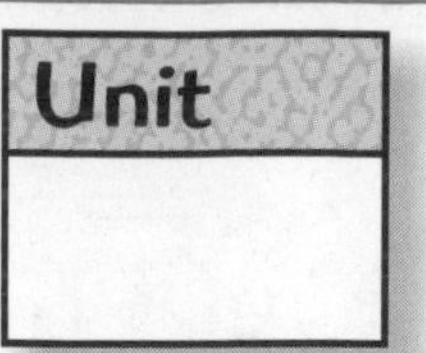

$$\begin{array}{r} 153 \\ -\ 28 \\ \hline \end{array}$$

3. 2 quarters = _____ dimes

2 quarters = _____ nickels

4 quarters = _____ dimes

4 quarters = _____ nickels

6 quarters = _____ dimes

6 quarters = _____ nickels

Tell a partner about the pattern you see.

4. 13 calculators. 2 children for each calculator. How many children?

_____ children

5. Use Ⓝ, Ⓓ, Ⓠ, and $1 to show $1.85 two ways.

6. Have a $10 bill. Spend $8.90. How much change?

______

Date Time

Counting Bills

1.     = $______

2.   = $______

3.     = $______

4.     
   = $______

5.    
   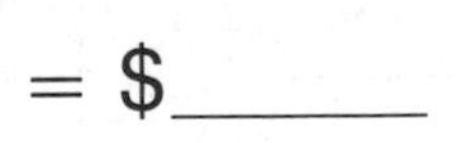 = $______

6.     
     = $ ______

7. Write the amounts in Problems 1–6 in order, from smallest to largest.

$______ $______ $______ $______ $______ $______

Date Time

Math Boxes 10.9

1. Cross out the names that do not belong.

1¢

one cent, $0.01, Ⓟ,

$\frac{1}{10}$ of a dime, $\frac{1}{10}$ of a dollar,

$\frac{1}{100}$ of a dollar, $0.10,

$\frac{1}{25}$ of a quarter

2. In 2,304 the value of

2 is ______.

3 is ______.

0 is ______.

4 is ______.

3. Write the amounts.

Five thousand six hundred eight

______

Two hundred sixteen dollars and sixty-eight cents ______

Three hundred nine dollars and five cents ______

4. Write the number that is

	10 more	100 less
368	______	______
4,789	______	______
40,870	______	______
1,999	______	______

5. Color $\frac{6}{8}$ of the circle.

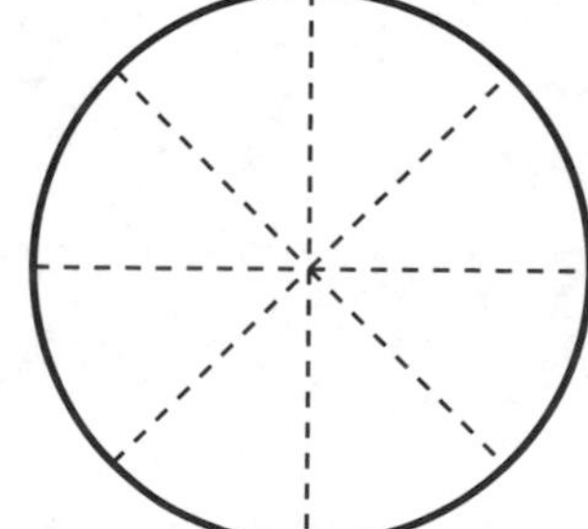

What fraction of the circle is not colored?

______

What is another name for that fraction? ______

6. Trade first. Then subtract.

$$\begin{array}{r} \$5.44 \\ -\ \$0.29 \\ \hline \end{array} \qquad \begin{array}{r} \$5.44 \\ -\ \$3.29 \\ \hline \end{array}$$

Date Time

Place Value

1. Match names.

A. 5 ones ______ 50

B. 5 tens ______ 500

C. 5 hundreds ______ 50,000

D. 5 thousands __A__ 5

E. 5 ten-thousands ______ 5,000

Fill in the blanks. Write ones, tens, hundreds, thousands, or ten-thousands.

2. The 8 in 74,863 stands for 8 ____________________.

3. The 6 in 35,926 stands for 6 ____________________.

4. The 2 in 2,785 stands for 2 ____________________.

5. The 5 in 58,047 stands for 5 ____________________.

6. The 0 in 13,409 stands for 0 ____________________.

Continue.

7. 364; 365; 366; ________; ________; ________

8. 1,796; 1,797; 1,798; ________; ________; ________

9. 996; 997; 998; ________; ________; ________

10. 1,996; 1,997; 1,998; ________; ________; ________

11. 9,996; 9,997; 9,998; ________; ________; ________

Date Time

Math Boxes 10.10

1. You have 21 pennies to share equally among 3 children.

How many pennies does each child get?

_______ pennies

How many left over?

_______ pennies

2. In 4,057 the value of

4 is _______.

0 is _______.

5 is _______.

7 is _______.

3. Solve. Use partial sums.

$$\begin{array}{r} \$3.74 \\ +\ \$0.27 \\ \hline \end{array} \qquad \begin{array}{r} \$3.74 \\ +\ \$4.27 \\ \hline \end{array}$$

4.

Rule
5Ⓝ = 1Ⓠ

Ⓝ	Ⓠ
	3
	4
30	
	10
	20

5. I had a 10-dollar bill. I spent $5.23. How much change did I receive?

____

6.

−10

1,523 → ☐ → ☐ → 1,713 → ☐ → ☐

Date Time

Parentheses Puzzles

Parentheses can make a big difference in a problem.

Example

$15 - 5 + 3 = ?$

$(15 - 5) + 3 = (10) + 3 = 13$; but

$15 - (5 + 3) = 15 - (8) = 7$

Solve problems containing parentheses.

1. $7 + (8 - 3) =$ _____

2. $(4 + 11) - 6 =$ _____

3. $8 + (13 - 9) =$ _____

4. _____ $= (12 + 8) - 16$

5. $140 - (20 + 80) =$ _____

6. _____ $= (30 + 40) - 70$

Put in parentheses to solve the puzzles.

7. $12 - 4 + 6 = 14$

8. $15 - 9 - 4 = 10$

9. $140 - 60 + 30 = 110$

10. $500 = 400 - 100 + 200$

11. $3 \times 2 + 5 = 11$

12. $2 \times 5 - 5 = 0$

Date Time

Math Boxes 10.11

1. Circle $\frac{3}{18}$.

What fraction of dots is not circled?

_____

2. Jordan spent $6.37 on a book and $1.23 on a magazine. How much did he spend altogether? First, estimate the costs and total.

_____ + _____ = _____

Then use partial sums and solve. _____

3. Write 5 names for $0.75.

____

____

____

____

____

4. You have 17 pieces of gum to share equally. If each child gets 4 pieces, how many children are sharing?

_____ children

How many pieces of gum are left over?

_____ pieces of gum

5. Write the number that is 100 more than

542 _____

837 _____

5,641 _____

9,863 _____

6. Use each digit once. Write the largest and smallest numbers.

3 1 4 6

Largest number: _____

Smallest number: _____

Read each of your numbers to a partner.

Date Time

Math Boxes 10.12

1. There are 3 drink boxes per pack. How many packs are needed to serve 25 second graders and 2 teachers one drink box each? Draw an array.

_______ packs are needed.

2. Count by 100s.

3,229; _______; _______;

_______; 3,629; _______;

_______; _______; _______

3. How many dimes in $1.00? _____

1 dime = _______ of $1.00.

How many cm in 1 meter? _____

1 cm = _______ of a meter.

How many inches in 2 feet?

_____

1 in. = _______ of 2 ft

4. Use each digit once. Write the largest and smallest numbers.

7 9 5 2

Largest number: _______

Smallest number: _______

Read each of your numbers to a partner.

5. Draw the hour and minute hands to show the time 20 minutes later than 6:15.

What time does the clock show now?

_____ : _____

6. If 25¢ is ONE,

what is 5¢? _______

what is 50¢? _______

If $2.00 is ONE,

what is 50¢? _______

what is $4.00? _______

Date Time

Art Supply Poster

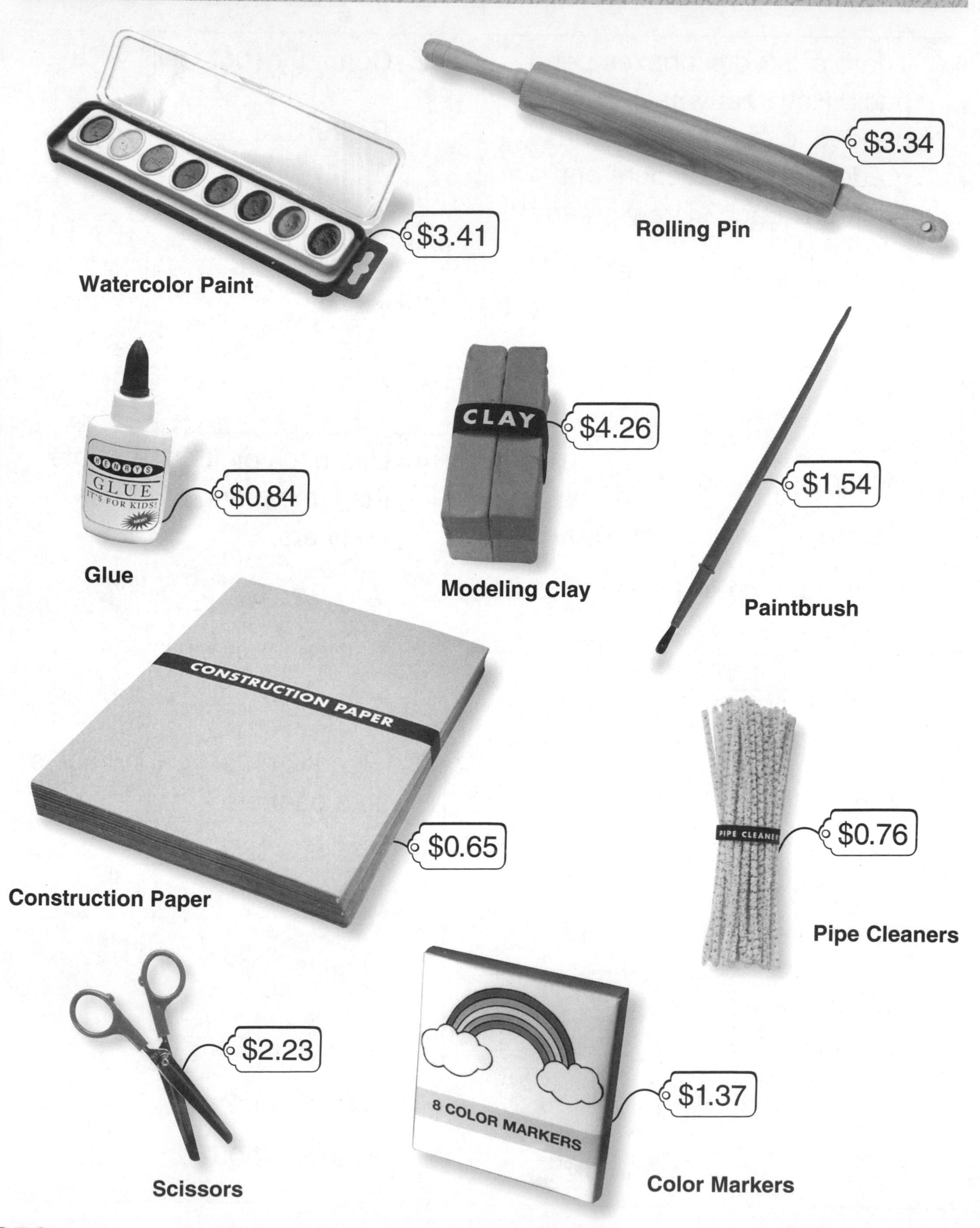

Use with Lesson 11.1.

Date Time

Buying Art Supplies

Estimate the total cost for each pair of items.

Write your estimate in the answer space.

If your estimate is *$3.00 or more,* add to find the total cost.

If your estimate is *less than $3.00,* do not find the total cost.

1. pipe cleaners and watercolors	2. clay and construction paper	3. paintbrush and scissors
Estimated Cost: ___ Total Cost: ___	Estimated Cost: ___ Total Cost: ___	Estimated Cost: ___ Total Cost: ___
4. glue and construction paper	**5.** markers and glue	**6.** clay and rolling pin
Estimated Cost: ___ Total Cost: ___	Estimated Cost: ___ Total Cost: ___	Estimated Cost: ___ Total Cost: ___

Date Time

Math Boxes 11.1

1. Solve.

Unit
apples

_____ = 4 + 8

_____ = 34 + 8

12 − 7 = _____

72 − 7 = _____

2. Write a number with 4 in the thousands place.

_____

What is the value of the digit 4 in your number?

_____

3. Draw 2 ways to show $\frac{2}{5}$.

4. Solve.

Unit

$$\begin{array}{r} 19 \\ +\ 8 \\ \hline \end{array} \qquad \begin{array}{r} 19 \\ 8 \\ +\ 5 \\ \hline \end{array} \qquad \begin{array}{r} 19 \\ 8 \\ 5 \\ +\ 16 \\ \hline \end{array}$$

5. Trade first, then subtract. Show your work.

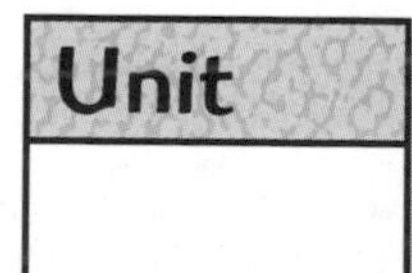

$$\begin{array}{r} 72 \\ -\ 35 \\ \hline \end{array}$$

6. Solve.

Unit
butterfly larvae

(15 − 5) + 3 = _____

15 − (5 + 3) = _____

17 + (9 − 4) = _____

(17 + 9) − 4 = _____

Date Time

Comparing Costs

Use the Art Supply Poster on journal page 272.
In Problems 1–6, circle the item that costs more.
Then find how much more.

1. glue or markers How much more? ________	**2.** rolling pin or scissors How much more? ________
3. pipe cleaners or paintbrush How much more? ________	**4.** construction paper or paintbrush How much more? ________
5. watercolors or markers How much more? ________	**6.** paintbrush or watercolors How much more? ________

7. You buy a pack of construction paper. You pay with a $1 bill.

Should you get more or less than 2 quarters in change? ________

8. You buy pipe cleaners. You pay with a $1 bill.

How much change should you get? ________

9. You buy a rolling pin. You pay with a $5 bill.

How much change should you get? ________

Date Time

Math Boxes 11.2

1. Kathleen wants to buy a sandwich and chips for \$2.25 and an apple for \$0.45. Is \$3.00 enough?

____

2. Write in dollars-and-cents notation:

two dimes _______

two pennies _______

two nickels _______

two quarters _______

3. Divide the rhombus into 4 equal parts.

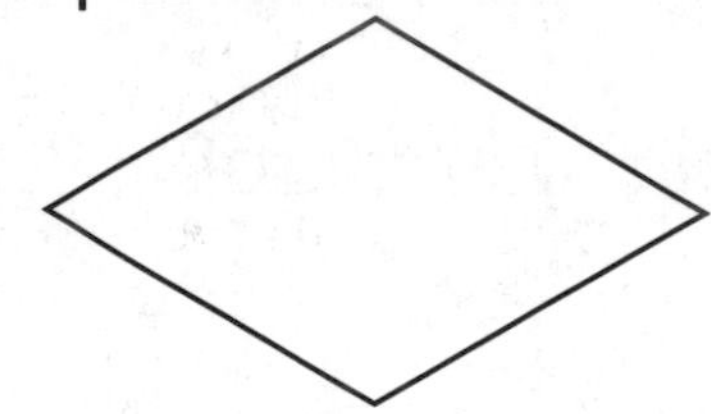

What fraction of the rhombus in each part?

____

4. __2__ nickels = 1 dime

1 nickel = __$\frac{1}{2}$__ dime

_____ inches = 1 foot

1 inch = _____ foot

_____ nickels = 1 quarter

1 nickel = _____ quarter

5. Count by 100s.

_______; _______; 2,748;

_______; _______; _______;

_______; _______

6. Complete the Fact Triangle. Write the fact family.

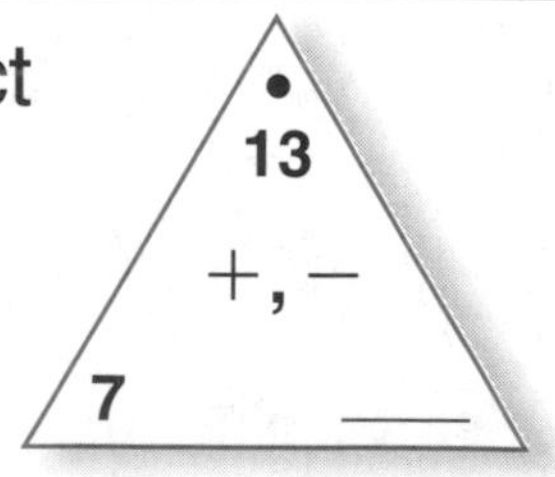

___ + ___ = ___

___ + ___ = ___

___ − ___ = ___

___ − ___ = ___

Date Time

Math Boxes 11.3

1. Circle the digit in the 1,000s place.

4, 6 9 4

2 9, 4 0 0

2 0, 0 0 4

5, 0 1 9

Read each number to a partner.

2. Tammy paid \$7.38 for her tape. Robin paid \$7.79. How much more did Robin pay?

______

3. Solve.

Unit

$(14 - 7) + 4 =$ ____

$14 - (7 + 4) =$ ____

$(12 - 7) + 5 =$ ____

$12 - (7 + 5) =$ ____

4. Circle the best answer.

A gallon has about:

4 quarts 4 cups

4 pounds

A baby weighs about:

70 pounds 7 pounds

7 ounces

5. If 10¢ is ONE,

what is 5¢? ____

what is 20¢? ____

If \$1.00 is ONE,

what is 25¢? ____

what is \$5? ____

6. Draw an 8-by-4 array.

How many in all? ____

Date Time

Multiplication Number Stories

1. 3 vans full of people. How many people in all?

Holds 10 people

vans	people per van	people in all

Answer: _____ people

Number model: _____ × _____ = _____

2. 4 insects on the flower. How many legs in all?

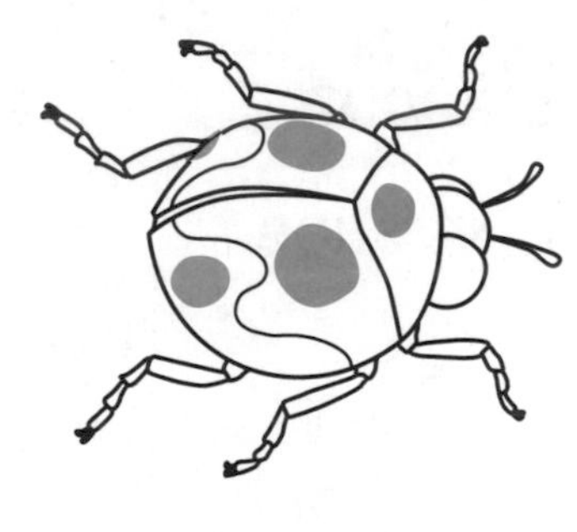

Has 6 legs

insects	legs per insect	legs in all

Answer: _____ legs

Number model: _____ × _____ = _____

3. 9 windows. How many panes in all?

Has 4 panes

windows	panes per window	panes in all

Answer: _____ panes

Number model: _____ × _____ = _____

Date Time

Multiplication Number Stories (cont.)

Use the pictures to make up two multiplication number stories.

Has 7 candles

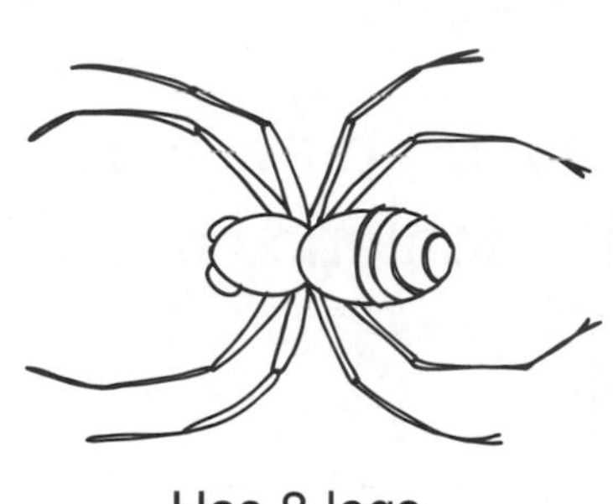

Has 8 legs

Has 5 players

For each story:

- Fill in the multiplication diagram.
- Draw a picture or array and find the answer.
- Fill in the number model.

4. ______________________

____

____

______	______ per ______	______ **in all**

Answer: ____

Number model: ____ × ____ = ____

5. ______________________

____

____

______	______ per ______	______ **in all**

Answer: ____

Number model: ____ × ____ = ____

Date Time

Division Number Stories

For each number story:

- Fill in the diagram.
- On a separate sheet of paper, draw a picture or array and find the answer. Complete the sentences.
- Fill in the number model.

1. The pet shop has 12 puppies in pens. There are 4 puppies in each pen. How many pens have puppies in them?

pens	puppies per pen	puppies in all

_____ pens have puppies in them. _____ puppies are left over.

Number model: _____ ÷ _____ → _____ R_____

2. Five children are playing a game with a deck of 30 cards. How many cards can the dealer give each player?

children	cards per child	cards in all

_____ cards to each player. _____ cards are left over.

Number model: _____ ÷ _____ → _____ R_____

3. Eight children share 18 toys equally. How many toys does each child get?

_____	_____ per _____	_____ in all

Each child gets _____ toys.

_____ toys are left over.

Number model: _____ ÷ _____ → _____ R_____

Date Time

Division Number Stories (cont.)

4. Tennis balls are sold 3 to a can. Rebecca buys 15 balls. How many cans is that?

cans	balls per can	balls in all

Rebecca buys _____ cans.

Number model: _____ ÷ _____ → _____ R_____

---

5. Seven friends share 24 marbles equally. How many marbles does each friend get?

_____	_____ per _____	_____ in all

Each friend gets_____ marbles. _____ marbles are left over.

Number model: _____ ÷ _____ → _____ R_____

---

6. Tina is storing 20 packages of seeds in boxes. Each box holds 6 packages. How many boxes does Tina need to store all the packages? (Be careful. Think!)

_____	_____ per _____	_____ in all

Tina needs _____ boxes.

Number model: _____ ÷ _____ → _____ R_____

Date Time

Math Boxes 11.4

1. 5 wagons. 4 wheels on each wagon. How many wheels?

_____ wheels

Fill in the diagram and write a number model.

wagons	wheels per wagon	wheels in all

_____ × _____ = _____

2. Sue spent $0.88 on glitter stickers and $0.23 on a large teddy bear sticker. How much did she spend?

Estimate:

_____ + _____ = _____

Answer:

_____

3. What is the value of the digit 4 in each number?

14 _____

142 _____

436 _____

4,678 _____

4. Count by 10s.

_____, 882, _____,

_____, _____, _____,

_____, _____

5. You buy some stickers for $1.89. Show 2 ways to pay. Use Ⓟ, Ⓝ, Ⓓ, Ⓠ, and [$1].

6. Solve.

Unit

300 + 200 + 40 + 18 = _____

40 + 20 + 19 = _____

_____ = 400 + 260 + 9

_____ = 500 + 130 + 27

Date Time

Multiplication Facts List

I am listing the times _____ facts.
If you are not sure of a fact, draw an array with Os or Xs.

2 × _____ = _____

3 × _____ = _____

4 × _____ = _____

5 × _____ = _____

6 × _____ = _____

7 × _____ = _____

8 × _____ = _____

9 × _____ = _____

10 × _____ = _____

Date Time

Using Arrays to Find Products

Draw an array to help you find each product.
Use Xs to draw your arrays.

1. $2 \times 4 =$ _____ X X X X X X X X	**2.** $4 \times 2 =$ _____
3. $6 \times 5 =$ _____	**4.** $5 \times 6 =$ _____
5. $5 \times 5 =$ _____	**6.** $2 \times 10 =$ _____

Challenge

7. $4 \times 15 =$ _____

Math Boxes 11.5

1. Each square has 4 corners. 16 corners in all. How many squares? _____ squares
Fill in the diagram and write a number model.

squares	*corners* per *square*	*corners* in all
?		*16*

_____ ÷ _____ → _____ R_____

2. 3 insects. 6 legs per insect. How many legs in all?

_____ legs

Fill in the diagram and write a number model.

insects	legs per insect	legs in all

_____ × _____ = _____

3. Write another name for each number.

50 tens = ____________

32 hundreds = ____________

6,240 = 624 ____________

12,000 = 12 ____________

4. A movie ticket costs \$2.50. Popcorn costs \$1.75. I have \$5.00. Is that enough to buy both?

_____

5. Write a number or a fraction.

There are _____ quarters in one dollar.

One quarter = _____ dollar.

There are _____ dimes in one dollar.

One dime = _____ dollar.

6. Draw a rectangle.
Make 2 sides $4\frac{1}{2}$ cm long.
Make the other 2 sides $2\frac{1}{2}$ cm long.

Date Time

Products Table

0×0 = **0**	0×1 =	0×2 =	0×3 =	0×4 =	0×5 =	0×6 =	0×7 =	0×8 =	0×9 =	0×10 =
1×0 =	1×1 = **1**	1×2 =	1×3 =	1×4 =	1×5 =	1×6 =	1×7 =	1×8 =	1×9 =	1×10 =
2×0 =	2×1 =	2×2 = **4**	2×3 =	2×4 =	2×5 =	2×6 =	2×7 =	2×8 =	2×9 =	2×10 =
3×0 =	3×1 =	3×2 =	3×3 = **9**	3×4 =	3×5 =	3×6 =	3×7 =	3×8 =	3×9 =	3×10 =
4×0 =	4×1 =	4×2 =	4×3 =	4×4 = **16**	4×5 =	4×6 =	4×7 =	4×8 =	4×9 =	4×10 =
5×0 =	5×1 =	5×2 =	5×3 =	5×4 =	5×5 = **25**	5×6 =	5×7 =	5×8 =	5×9 =	5×10 =
6×0 =	6×1 =	6×2 =	6×3 =	6×4 =	6×5 =	6×6 = **36**	6×7 =	6×8 =	6×9 =	6×10 =
7×0 =	7×1 =	7×2 =	7×3 =	7×4 =	7×5 =	7×6 =	7×7 = **49**	7×8 =	7×9 =	7×10 =
8×0 =	8×1 =	8×2 =	8×3 =	8×4 =	8×5 =	8×6 =	8×7 =	8×8 = **64**	8×9 =	8×10 =
9×0 =	9×1 =	9×2 =	9×3 =	9×4 =	9×5 =	9×6 =	9×7 =	9×8 =	9×9 = **81**	9×10 =
10×0 =	10×1 =	10×2 =	10×3 =	10×4 =	10×5 =	10×6 =	10×7 =	10×8 =	10×9 =	10×10 = **100**

Date Time

Multiplication with 2, 5, and 10

1. double 5 = _____ _____ = double 6

   $2 \times 3 =$ _____ _____ $= 2 \times 0$

   $6 \times 2 =$ _____ _____ $= 1 \times 2$

2. 4 nickels = _____ cents _____ $= 0 \times 5$

   $4 \times 5 =$ _____ _____ cents = 8 nickels

   $5 \times 1 =$ _____ _____ $= 5 \times 5$

3. 3 dimes = _____ cents _____ $= 10 \times 1$

   $3 \times 10 =$ _____ _____ cents = 6 dimes

   $10 \times 2 =$ _____ _____ $= 0 \times 10$

4. Write the turn-around fact for each fact.

   $2 \times 7 = 14$ 7 × 2 = 14

   $5 \times 6 = 30$ _____ × _____ = _____

   $10 \times 9 = 90$ _____ × _____ = _____

   $9 \times 2 = 18$ _____ × _____ = _____

   $5 \times 7 = 35$ _____ × _____ = _____

   $10 \times 8 = 80$ _____ × _____ = _____

Date Time

Math Boxes 11.6

1. Use an inch ruler to find the perimeter of the hexagon.

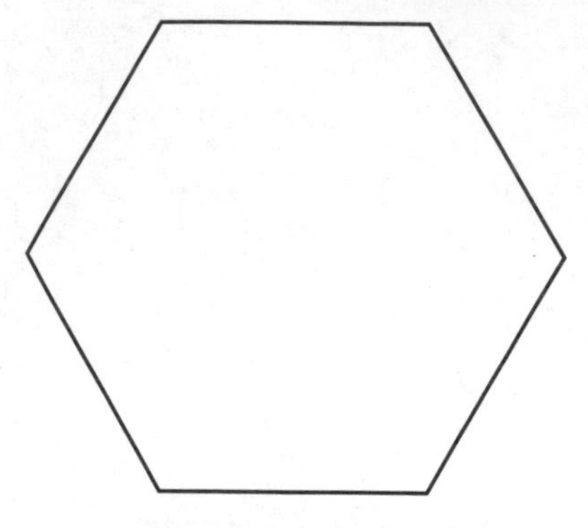

The perimeter is _____ inches.

2. \$24.06 \$9.99 \$14.98 \$19.99 \$29.83

The median cost of a video is

\$_______.

3. Add parentheses to make each number model true.

16 − 9 + 7 = 14

16 − 9 + 7 = 0

29 − 9 + 7 = 27

29 − 9 + 7 = 13

4. There are 20 dolls on 4 shelves. Each shelf has the same number of dolls. How many dolls on each shelf?

_____ dolls with

_____ dolls left over

Write a number model.

____ ÷ ____ → ____ R____

5. Fill in the table.

Rule
×10

in	out
1	
6	
10	
	80
	250

6. A baseball costs \$3.69. A yo-yo costs \$1.49. You buy both.

Estimate the cost:

______ + ______ = ______

Actual cost:

______

Date Time

Multiplication/Division Fact Families

Write the fact family for each Fact Triangle.

1.

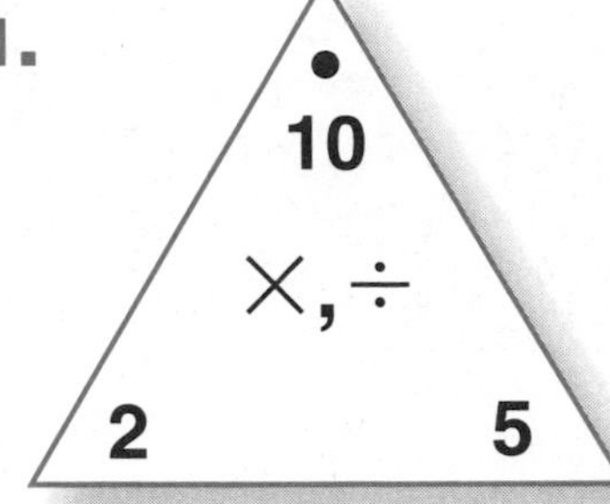

5 × 2 = 10

___ × ___ = ___

10 ÷ 2 = 5

___ ÷ ___ = ___

2.

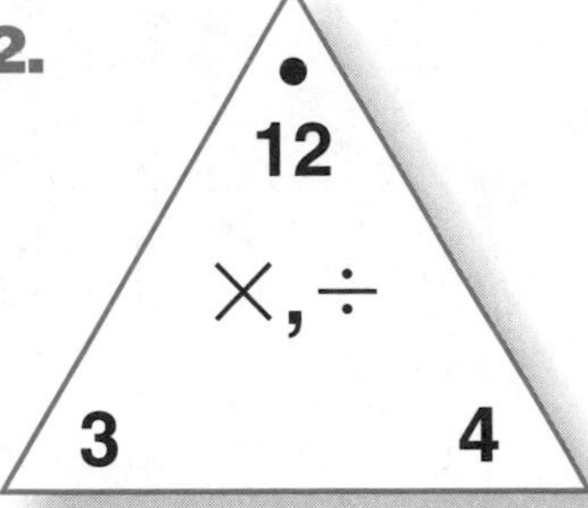

___ × ___ = ___

___ × ___ = ___

___ ÷ ___ = ___

___ ÷ ___ = ___

3.

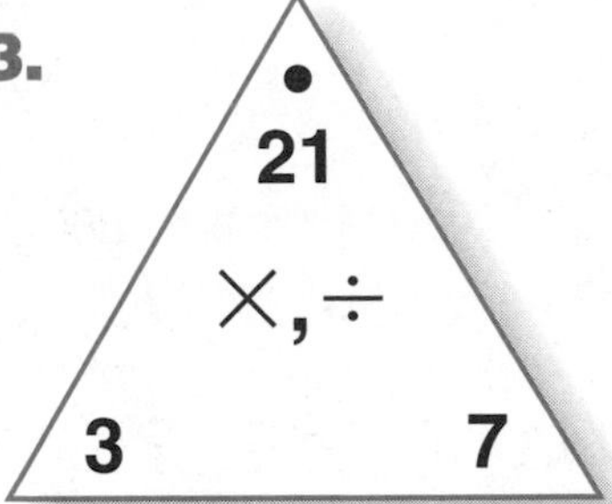

___ × ___ = ___

___ × ___ = ___

___ ÷ ___ = ___

___ ÷ ___ = ___

4.

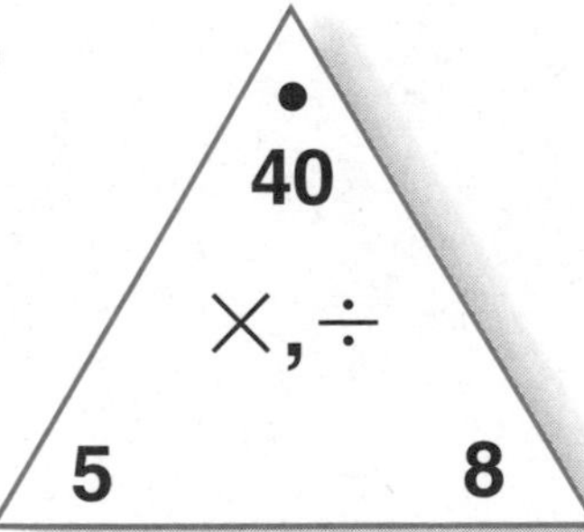

___ × ___ = ___

___ × ___ = ___

___ ÷ ___ = ___

___ ÷ ___ = ___

5.

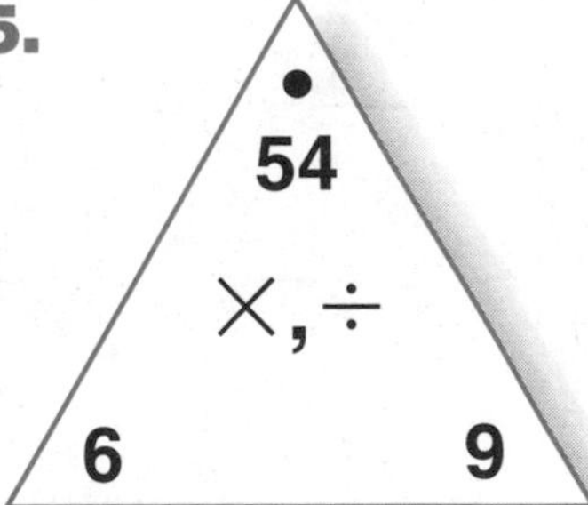

___ × ___ = ___

___ × ___ = ___

___ ÷ ___ = ___

___ ÷ ___ = ___

6.

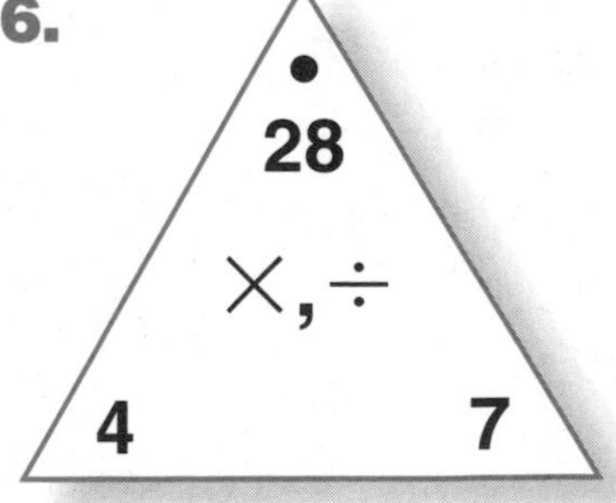

___ × ___ = ___

___ × ___ = ___

___ ÷ ___ = ___

___ ÷ ___ = ___

Date Time

Multiplication and Division with 2, 5, and 10

1. double 8 = _____ _____ = double 9

$2 \times 4 =$ _____ _____ $= 2 \times 1$

$7 \times 2 =$ _____ _____ $= 0 \times 2$

2. 40 cents = _____ nickels $5 \div 1 =$ _____

$15 \div 5 =$ _____ _____ nickels = 25 cents

3. 40 cents = _____ dimes _____ $\div 10 = 6$

$20 \div 10 =$ _____ _____ dimes = 90 cents

4. For each multiplication fact, give two division facts in the same fact family.

$2 \times 6 = 12$ 12 ÷ 2 = 6 12 ÷ 6 = 2

$5 \times 9 = 45$ _____ ÷ _____ = _____ _____ ÷ _____ = _____

$10 \times 4 = 40$ _____ ÷ _____ = _____ _____ ÷ _____ = _____

$3 \times 2 = 6$ _____ ÷ _____ = _____ _____ ÷ _____ = _____

$8 \times 5 = 40$ _____ ÷ _____ = _____ _____ ÷ _____ = _____

$5 \times 10 = 50$ _____ ÷ _____ = _____ _____ ÷ _____ = _____

Date Time

Math Boxes 11.7

1. Mrs. Bell had 30 pennies. She gave $\frac{1}{3}$ of the pennies to Max and $\frac{1}{2}$ of the pennies to Julie.

   Max received _____ pennies.

   Julie received _____ pennies.

   How many pennies did Mrs. Bell have left? _____ pennies

2. Write >, <, or =.

   70 + 39 _____ 59 + 60

   98 − 70 _____ 62 − 40

   156 − 90 _____ 26 + 40

3. Leah swam 10 meters. Andrea swam 6 times as far. How far did Andrea swim?

   _____ meters

   Write the number model.

   ____

4. Use your calculator. Enter the problems. Write the answers in dollars and cents.

   64¢ + \$1.73 = \$_______

   85¢ + 53¢ = \$_______

   \$2.08 + \$5.01 = \$_______

   37¢ + 26¢ = \$_______

5. Estimate, then solve.

   $$\begin{array}{r} 262 \\ -\ 139 \\ \hline \end{array}$$

   Estimate:

   _____ − _____ = _____

   Answer: _____

6. Multiply. If you need help, make arrays.

Unit

   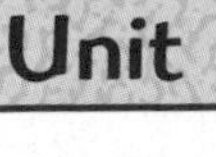

   $4 \times 5 =$ _____

   $6 \times 2 =$ _____

   $1 \times 10 =$ _____

Date Time

Beat the Calculator

Materials ❑ calculator

Players 3 (Caller, Brain, and Calculator)

Directions

1. The Caller reads fact problems from the Brain's journal—in the order listed on the next page.
2. The Brain solves each problem and says the answer.
3. While the Brain is working on the answer, the Calculator solves each problem using a calculator and says the answer.
4. If the Brain beats the Calculator, the Caller makes a check mark next to the fact in the Brain's journal.

Date Time

Beat the Calculator (cont.)

✓	✓	✓	Fact Problem
			2 × 4 = ____
			3 × 5 = ____
			2 × 2 = ____
			4 × 3 = ____
			5 × 5 = ____
			6 × 2 = ____
			6 × 5 = ____
			3 × 3 = ____
			4 × 5 = ____
			3 × 6 = ____

✓	✓	✓	Fact Problem
			7 × 3 = ____
			5 × 2 = ____
			6 × 4 = ____
			2 × 7 = ____
			3 × 2 = ____
			4 × 4 = ____
			4 × 1 = ____
			4 × 7 = ____
			7 × 5 = ____
			0 × 2 = ____

Date Time

Days of Rain or Snow in 1 Year (12 Months)

Use with Lesson 11.9.

Date Time

Days of Rain or Snow in 1 Year (cont.)

Rain and snow are called **precipitation**.

1. About how many days per year does it rain or snow in these cities?

City	Days of Precipitation	City	Days of Precipitation
Los Angeles	about _____	New York City	about _____
San Francisco	about _____	San Juan	about _____
Houston	about _____	Juneau	about _____
Atlanta	about _____		

2. Use the information above for problems 2–6.

	Name of City	Number of Days
Most days of precipitation:	_______________	about _____
Fewest days of precipitation:	_______________	about _____

Middle value: about _____ days

Range: about _____ days

Name of Place

3. Atlanta has about 55 more days of rain or snow than _______________.

4. New York City has about 80 fewer days of rain or snow than _______________.

5. Juneau has about twice as many days of rain or snow as _______________.

6. San Francisco has about $\frac{1}{2}$ as many days of rain or snow as _______________.

Date Time

Math Boxes 11.8

1. Solve.

$3 \times 0 =$ _____

$17 \times 0 =$ _____

$100 \times 0 =$ _____

$4 \times 1 =$ _____

$8 \times 1 =$ _____

$129 \times 1 =$ _____

2. Write the fact family.

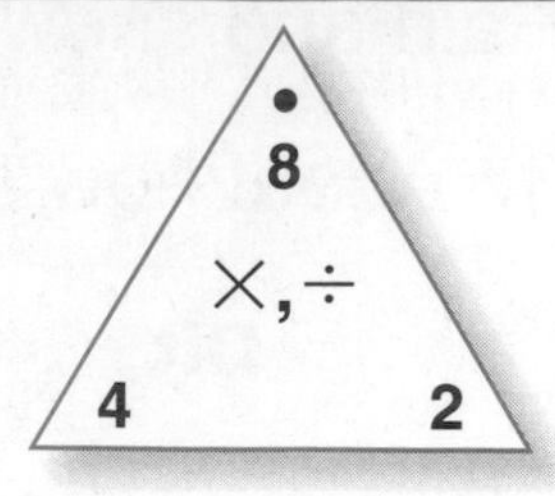

_____ × _____ = _____

_____ × _____ = _____

_____ ÷ _____ = _____

_____ ÷ _____ = _____

3. 18 children. 2 children per bench. How many benches?

_____ benches

Write the number model.

_____ ÷ _____ → _____ R_____

4. Solve.

$$\begin{array}{r} \$4.62 \\ - \ \$2.37 \\ \hline \end{array}$$

5. Write the number. Use your Place-Value Book if you need to.

3 tens = 30

33 tens = _____

333 tens = _____

6. Add parentheses to make the number models true.

Unit
children

$18 - 13 - 4 = 9$

$18 - 13 - 4 = 1$

$27 - 6 + 10 = 31$

$4 \times 2 + 3 = 20$

Date Time

Math Boxes 11.9

1. Draw 2 ways to show $\frac{1}{3}$.

2. Draw a shape with an area of 12 square centimeters.

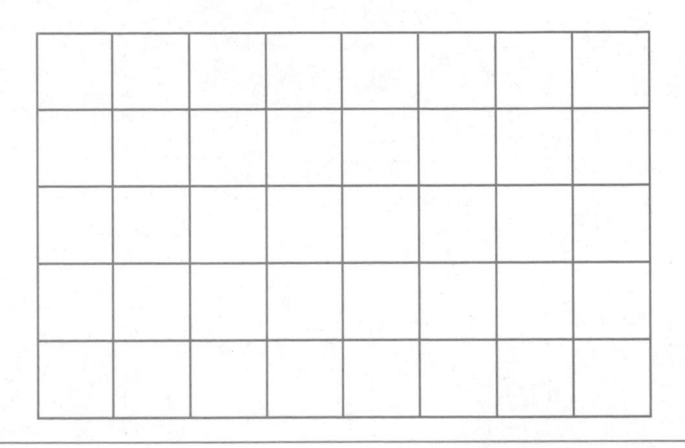

3. $\frac{1}{4}$ of the pennies is 6.

How many pennies in all?

_____ pennies

Use counters to help.

4. Solve.

Unit

$3 \times 6 =$ _____

$6 \times 3 =$ _____

$5 \times 2 =$ _____

$2 \times 5 =$ _____

5. Complete the Fact Triangle. Write the fact family.

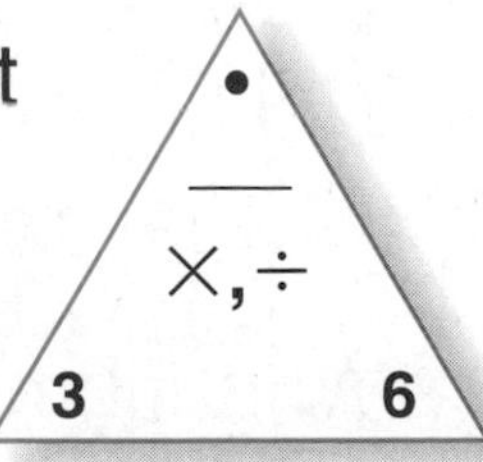

_____ × _____ = _____

_____ × _____ = _____

_____ ÷ _____ = _____

_____ ÷ _____ = _____

6. Solve.

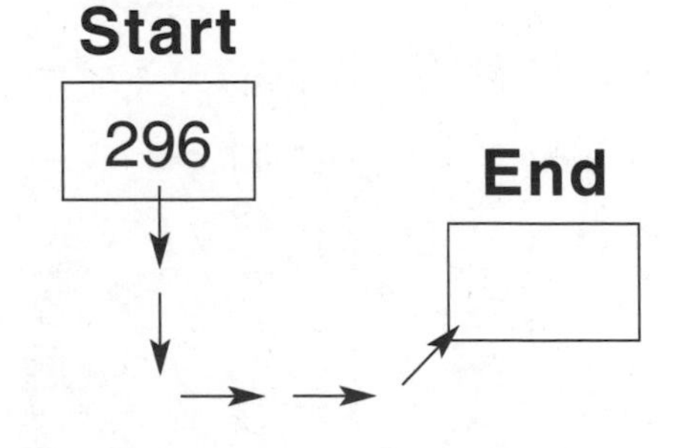

Date Time

Math Boxes 11.10

1. Solve.

$0 = 6 \times$ _____

$0 \times 90 =$ _____

$4 = 1 \times$ _____

$25 \times 0 =$ _____

$72 = 72 \times$ _____

$5{,}000 \times 0 =$ _____

2. Complete the Fact Triangle. Write the fact family.

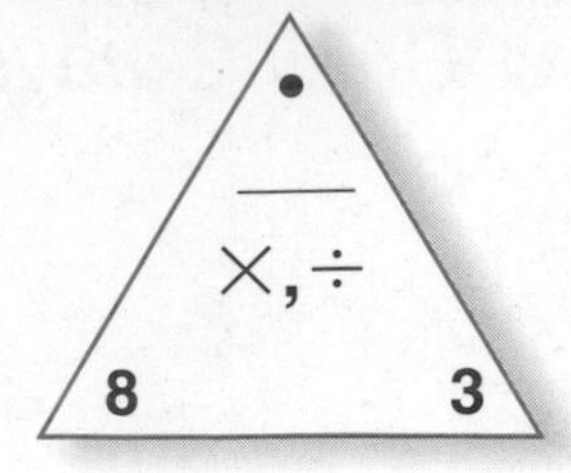

_____ × _____ = _____

_____ × _____ = _____

_____ ÷ _____ = _____

_____ ÷ _____ = _____

3. 15 baseball cards are shared equally among 4 children. How many cards does each child get? _____ cards

Write the number model.

_____ ÷ _____ → _____ R_____

4. A number has:

7 thousands

8 tens

5 ten-thousands

1 one

0 hundreds

Write the number. __________

5. The second grade play started at 1:45. It lasted 45 minutes. Show the time it ended.

What time did the play end?

_____:_____

6. What number is twice as much as 30? _____

What number is half as much as 30? _____

What number is 10 less than 30? _____

What number is 5 more than 30? _____

 Use with Lesson 11.10.

Date Time

Math Boxes 12.1

1. Draw a shape with a perimeter of 14 cm.

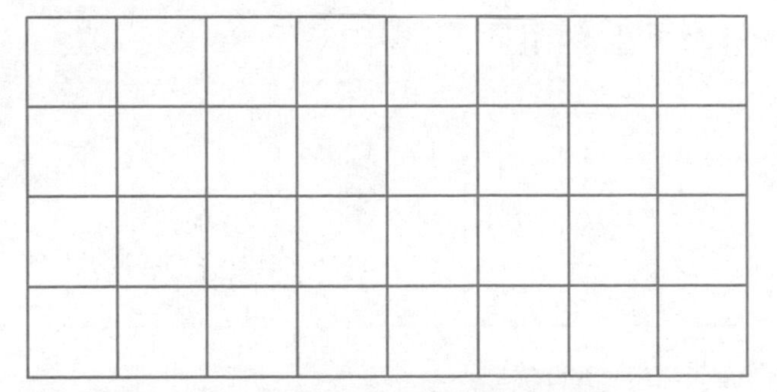

The area of the shape is

_____ square centimeters.

2. I spent \$4.22 at the store and gave the cashier a \$10 bill. How much change should I get?

\$_______

3. Solve.

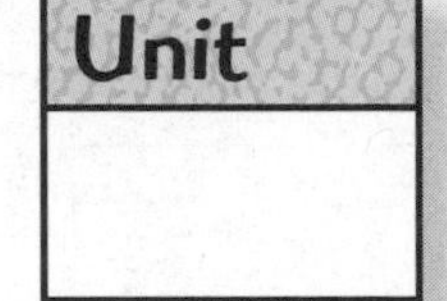

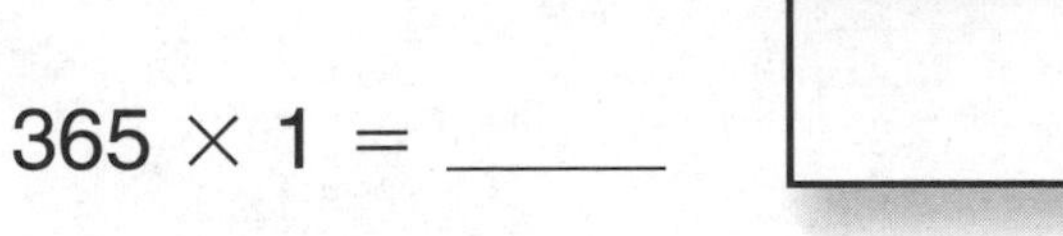

365 × 1 = _____

365 + 1 = _____

_____ = 444 × 0

_____ = 444 + 0

4. Draw a line segment 3 cm long.

Draw a second line segment 4 cm longer than the first.

Draw a third line segment twice as long as the first.

5. Complete the Fact Triangle. Write the fact family.

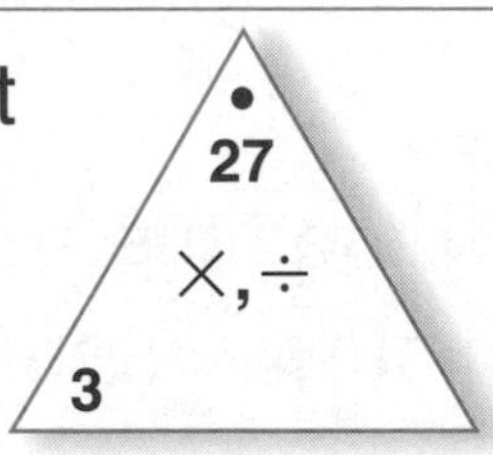

____ × ____ = ____

____ × ____ = ____

____ ÷ ____ = ____

____ ÷ ____ = ____

6. 8 flower boxes. 4 plants in each box. How many plants?

_____ plants

Write a number model.

____ × ____ = ____

Date Time

Time Before and After

1. It is:

Show the time
20 minutes later.

What time is it?

_____ : _____

2. It is:

Show the time
15 minutes earlier.

What time is it?

_____ : _____

3. It is:

Show the time
35 minutes later.

What time is it?

_____ : _____

4. You pick a time. Draw the hands on the clock.

It is:

Show the time
50 minutes later.

What time is it?

_____ : _____

Date Time

Many Names for Times

What time does each clock show? Fill in the ovals next to the correct names.

Example ● a quarter-past 1 ● 15 minutes after 1

O two ten O 5 minutes after 3

● one fifteen

---

1. O seven fifteen O a quarter-to 7

O a quarter-to 8 O a quarter-past 8

O a quarter-past 7

---

2. O half-past 10 O eleven thirty

O half-past 11 O 30 minutes after 10

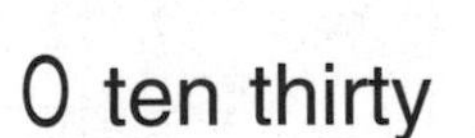

O ten thirty

---

3. O a quarter-past 5 O a quarter-to 6

O a quarter-to 5 O six fifteen

O five forty-five

---

4. O nine forty O 20 to 8

O 20 to 9 O eight forty

O 40 minutes to 9

Date Time

Addition Practice

Add. Use your favorite addition strategy. Be sure to check your answers. Use the answers at the bottom of journal page 303.

Unit

Easy

1. 53 + 45

Answer

2. $\begin{array}{r} 357 \\ +\ 201 \\ \hline \end{array}$

Answer

3. 64 + 26

Answer

Harder

4. 263 + 17

Answer

5. $\begin{array}{r} 36 \\ +\ 48 \\ \hline \end{array}$

Answer

6. $\begin{array}{r} 456 \\ +\ 275 \\ \hline \end{array}$

Answer

7. $\begin{array}{r} 5{,}174 \\ +\ 2{,}387 \\ \hline \end{array}$

Answer

8. 9,435 + 265

Answer

9. $\begin{array}{r} 7{,}496 \\ +\ 4{,}835 \\ \hline \end{array}$

Answer

Answers for page 303:

1. 44
2. 531
3. 153
4. 38
5. 478
6. 309
7. 468
8. 8,572
9. 99

Use with Lesson 12.2.

Date Time

Subtraction Practice

Subtract. Use your favorite subtraction strategy. Be sure to check your answers. Use the answers at the bottom of journal page 302.

Unit

Easy

1. 68 − 24 Answer	2. 563 − 32 Answer	3. 486 − 333 Answer

Harder

4. 65 − 27 Answer	5. 784 − 306 Answer	6. 555 − 246 Answer
7. 506 − 38 Answer	8. 9,006 − 434 Answer	9. 233 − 134 Answer

Answers for page 302:

1. 98
2. 558
3. 90
4. 280
5. 84
6. 731
7. 7,561
8. 9,700
9. 12,331

Date Time

Math Boxes 12.2

1. ______ months = 1 year

______ months = $\frac{1}{2}$ year

______ months = $\frac{1}{4}$ year

______ months = 2 years

2. Add.

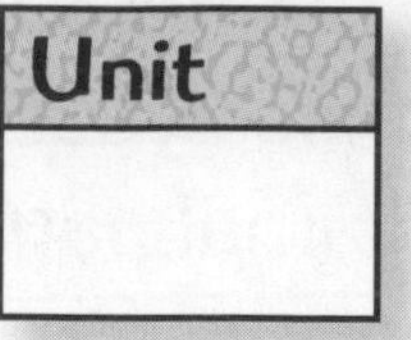

12,469 + 10 = ________

12,469 + 100 = ________

12,469 + 1,000 = ________

12,469 + 10,000 = ________

3. 20 campers divided equally among 5 tents. How many campers in each tent?

______ campers

Write a number model.

____

4. Divide into:

halves fourths

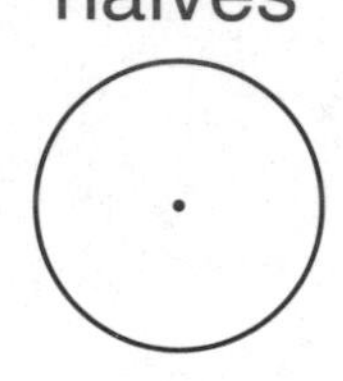

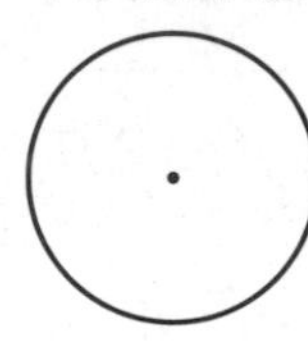

Write >, <, or =.

$\frac{1}{2}$ ______ $\frac{1}{4}$ $\frac{2}{4}$ ______ $\frac{1}{2}$

$\frac{1}{2}$ ______ $\frac{3}{4}$

5. Use your calculator to find the total.

[$1] [$1] [$1] [$1] = $______

Ⓠ Ⓠ Ⓠ = $______

Ⓓ Ⓓ Ⓓ Ⓓ Ⓓ = $______

Ⓝ Ⓝ Ⓝ Ⓝ Ⓝ Ⓝ Ⓝ = $______

Total = $______

6. I know: 17 − 9 = ______.

This helps me know:

37 − 9 = ______

177 − 9 = ______

687 − 9 = ______

Date Time

Math Boxes 12.3

1.

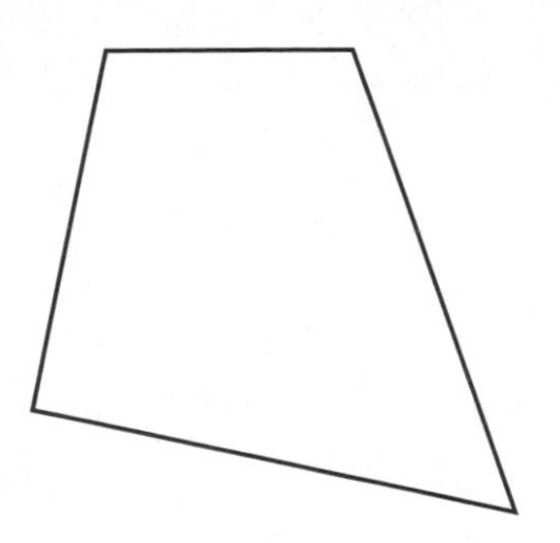

The perimeter is about

_____ cm.

2. Complete the Fact Triangle. Write the fact family.

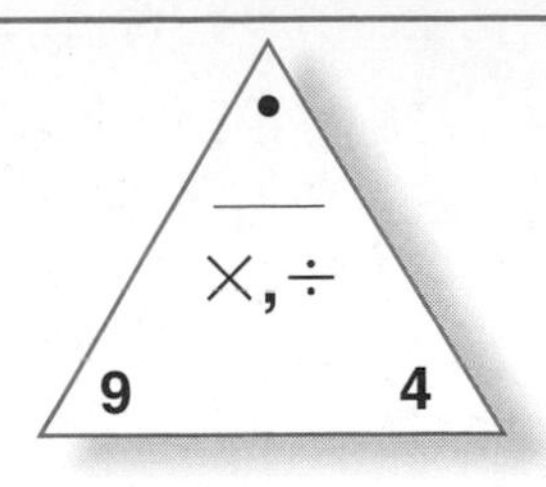

___ × ___ = ___

___ × ___ = ___

___ ÷ ___ = ___

___ ÷ ___ = ___

3. Keith left school at 3:25 P.M. He walked home in 15 minutes. What time did he arrive at home?

_____ : _____ P.M.

4. Solve.

Unit

$$\begin{array}{r} 687 \\ -\ 409 \\ \hline \end{array} \qquad \begin{array}{r} 569 \\ -\ 372 \\ \hline \end{array}$$

5. Cross out names that don't belong.

6:15

six fifteen, quarter to 7,
a quarter past 6,
15 minutes before 6,
15 minutes after 6

6. What number is twice as much as 23? _____

What number is half as much as 90? _____

What number is 20 less than 48? _____

Date Time

Important Events in Communication

For each event below, make a dot on the timeline and write the letter for the event above the dot.

A telephone (1876)

B radio (1906)

C television (1926)

D telegraph (1837)

E CD player (1982)

F 3-D movies (1922)

G audio cassette (1963)

H phonograph (1877)

I personal computer (1974)

J movie machine (1894)

K copier (1937)

L video cassette (1969)

M typewriter (1867)

N FM radio (1933)

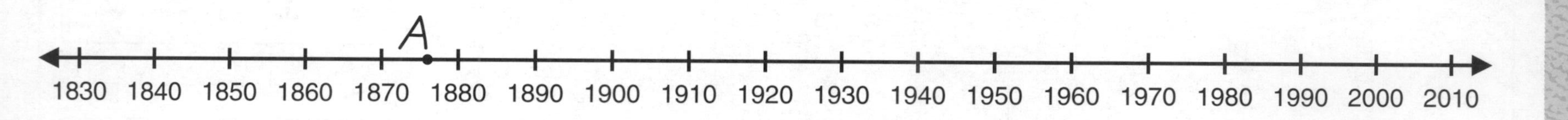

Interpreting a Timeline

1. What is the earliest invention on the timeline

on journal page 306? ____________________

What is the most recent invention? ____________________

For each pair of inventions:

- tell about how many decades there were between inventions.
- tell about how many years there were between inventions.

> ***Reminder:*** 1 decade is 10 years.
> 1 century is 100 years.
> 1 century is 10 decades.

2. typewriter and movie machine

about ______ decades about ______ years

3. phonograph and video cassette

about ______ decades about ______ years

4. telegraph and CD player

about ______ decades about ______ years

About how many years ago were these things invented?

5. CD player: about ______ years ago

6. FM radio: about ______ years ago

7. 3-D movies: about ______ years ago

8. typewriter: about ______ years ago

Date Time

Math Boxes 12.4

1. Put parentheses to make each number model true.

   $21 = 39 - 10 - 8$

   $4 \times 3 + 7 = 40$

   $3 \times 5 + 2 = 17$

2. Put the numbers in order.

   35, 64, 27, 86, 71

   ____, ____, ____, ____, ____

   The median is ____.

   The difference between the highest and lowest number (range) is ____.

3. Fill in the frames.

   ×10 −10

   2

4. In 43,692, the value of

   4 is ________.

   3 is ________.

   6 is ________.

   9 is ________.

   2 is ________.

5. Solve.

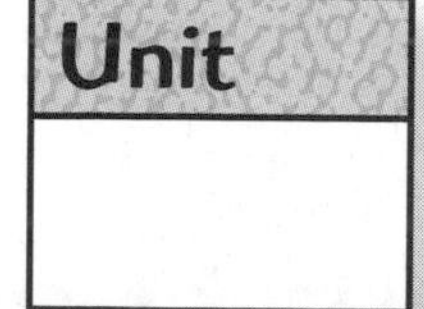

   $4 \times 0 =$ ____

   ____ $= 63 \times 1$

   $7 \times 10 =$ ____

   ____ $= 100 \times 3$

6. Julia has 9 teddy bears. Natalie has $\frac{1}{3}$ as many. How many does Natalie have?

   ____ teddy bears

   Draw a picture to help.

Date Time

Related Multiplication and Division Facts

Solve each multiplication fact.

Then use the three numbers to write two division facts.

1. 3 × 7 = 21

21 ÷ 7 = 3

21 ÷ 3 = 7

2. 3 × 8 = ____

____ ÷ ____ = ____

____ ÷ ____ = ____

3. 3 × 9 = ____

____ ÷ ____ = ____

____ ÷ ____ = ____

4. 4 × 7 = ____

____ ÷ ____ = ____

____ ÷ ____ = ____

5. 4 × 8 = ____

____ ÷ ____ = ____

____ ÷ ____ = ____

6. 4 × 9 = ____

____ ÷ ____ = ____

____ ÷ ____ = ____

7. 5 × 7 = ____

____ ÷ ____ = ____

____ ÷ ____ = ____

8. 5 × 8 = ____

____ ÷ ____ = ____

____ ÷ ____ = ____

Date Time

Related Multiplication and Division Facts (cont.)

9. $5 \times 9 =$ ____

____ ÷ ____ = ____

____ ÷ ____ = ____

10. $6 \times 7 =$ ____

____ ÷ ____ = ____

____ ÷ ____ = ____

11. $6 \times 8 =$ ____

____ ÷ ____ = ____

____ ÷ ____ = ____

12. $6 \times 9 =$ ____

____ ÷ ____ = ____

____ ÷ ____ = ____

13. $7 \times 7 =$ ____

____ ÷ ____ = ____

____ ÷ ____ = ____

14. $7 \times 8 =$ ____

____ ÷ ____ = ____

____ ÷ ____ = ____

15. $7 \times 9 =$ ____

____ ÷ ____ = ____

____ ÷ ____ = ____

16. $9 \times 9 =$ ____

____ ÷ ____ = ____

____ ÷ ____ = ____

Use with Lesson 12.5.

Date Time

Math Boxes 12.5

1.

Rule
$\times 4$

in	out
4	
9	
7	
	20
	40

2. Write the time in hours and minutes.

half-past six ____ : ____

quarter-past 4 ____ : ____

quarter-to 12 ____ : ____

twenty minutes to 2 ____ : ____

3. Jon's piano lesson started at a quarter-past four. It lasted 30 minutes. Show and write the time when he was finished.

____ : ____

4. Solve.

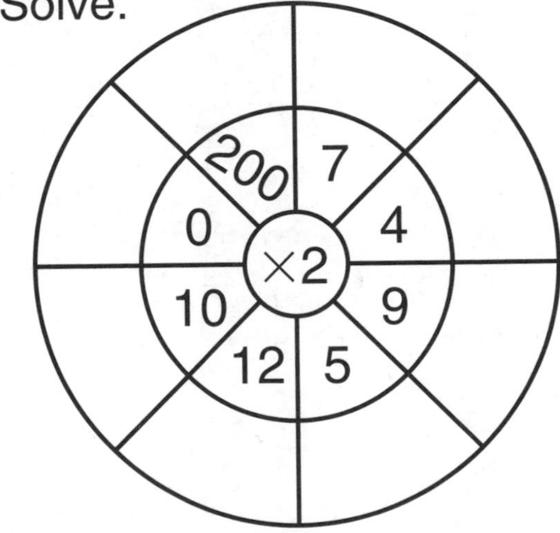

5. Write turnarounds and solve.

8×5 ____ $\times$ ____ $=$ ____

3×7 ____ $\times$ ____ $=$ ____

2×8 ____ $\times$ ____ $=$ ____

6×6 ____ $\times$ ____ $=$ ____

6. I want to buy a kite for $3.59 and string for $2.50. How many dollar bills do I need?

____ dollar bills

Date Time

Animal Bar Graph

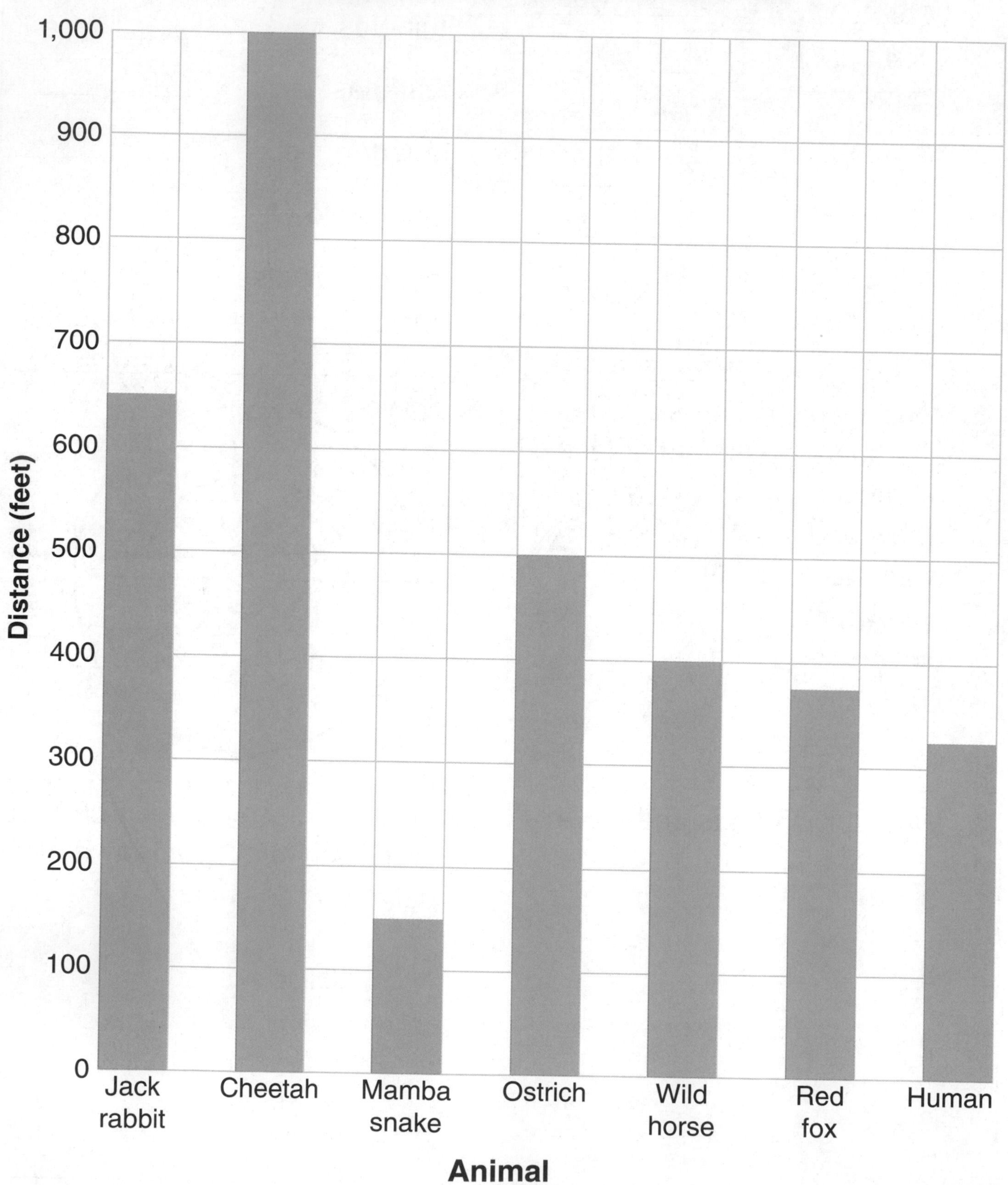

Use with Lesson 12.6.

Date Time

Interpreting an Animal Bar Graph

1. In the table, list the animals in order of distance covered in 10 seconds. List the animals from the greatest distance to the least distance.

2. Find the middle value of the distances. The middle value is also called the **median.**

   The median is _______ feet.

3. The longest distance is

   _______ feet.

   The shortest distance is

   _______ feet.

Distances Covered in 10 Seconds	
Animal	**Distance**
greatest: _______________	_______ ft
_______________	_______ ft
_______________	_______ ft
_______________	_______ ft
_______________	_______ ft
_______________	_______ ft
least: _______________	_______ ft

4. Fill in the comparison diagram with the longest distance and the shortest distance.

Quantity	
Quantity	**Difference**

5. Find the difference between the longest and shortest distances. The difference between the largest and smallest numbers in a data set is called the **range.**

   The range is _______ feet.

Distances

Distances athletic adults can travel in 10 seconds:

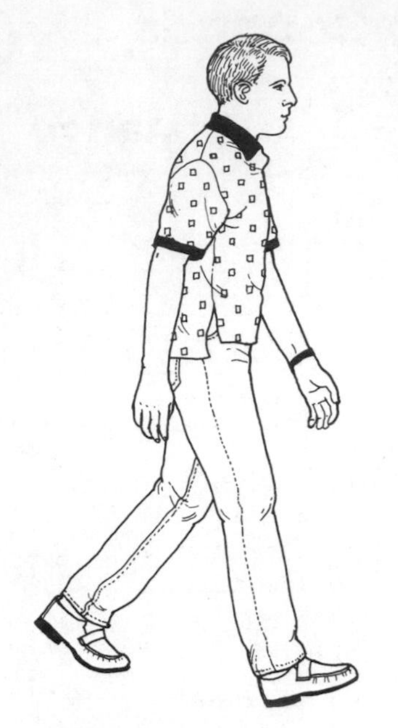

Walking 125 feet

Cross-country skiing 200 feet

Ice skating 450 feet

Running hurdles 275 feet

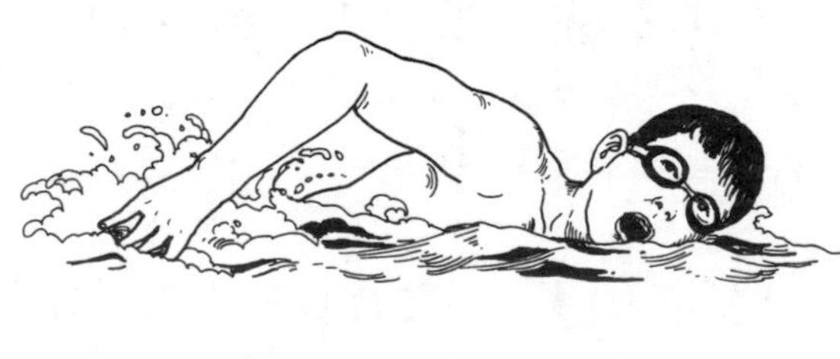

Swimming 75 feet

Running 325 feet

Date Time

Graphing Information

1. Complete the graph below with the information on journal page 314.

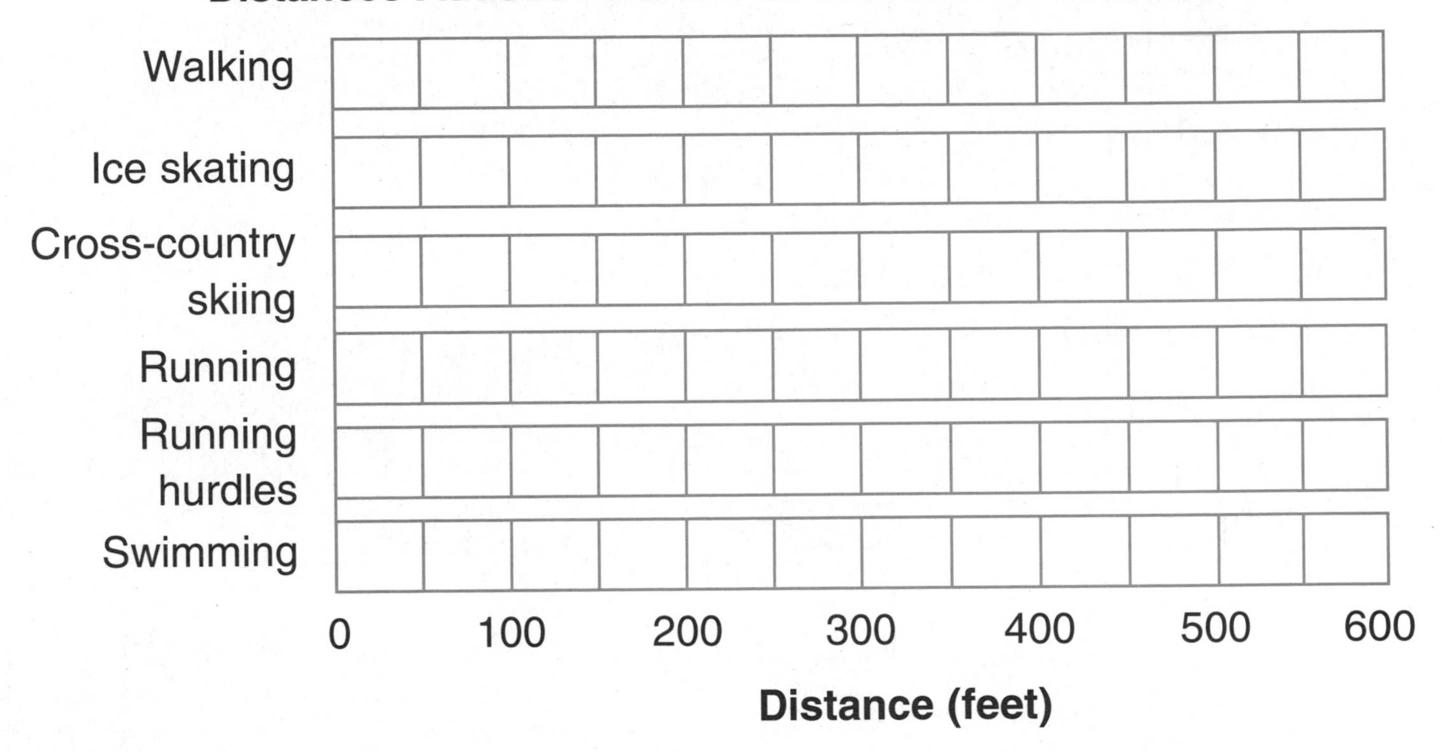

Interpret the graph.

2. The longest distance is _____ feet.

The shortest distance is _____ feet.

The difference between the longest distance and the shortest distance (range) is _____ feet.

3. What is a middle value of the distances on your graph? _____ feet

4. About how much distance can a cross-country skier cover:

in 20 seconds? _______ feet

in 30 seconds? _______ feet

in 1 minute? _______ feet

Date Time

Math Boxes 12.6

1. A shark swam 80 miles. A seal swam $\frac{1}{2}$ as far as the shark. How far did the seal swim?

_____ miles

A dolphin swam twice as far as the shark. How far did the dolphin swim?

_____ miles

2. Yes or no?

$8 \div 3 = 3 \div 8$ _____

$3 \times 9 = 9 \times 3$ _____

$100 \times 3 = 3 \times 100$ _____

$25 \div 5 = 5 \div 25$ _____

3. Solve.

Unit

$2{,}384 + 1 =$ _______

$2{,}384 + 10 =$ _______

$2{,}384 + 100 =$ _______

$2{,}384 + 1{,}000 =$ _______

$2{,}384 + 10{,}000 =$ _______

4. 1 hour = _____ minutes

$\frac{1}{2}$ hour = _____ minutes

$\frac{1}{4}$ hour = _____ minutes

$\frac{3}{4}$ hour = _____ minutes

$1\frac{1}{2}$ hours = _____ minutes

5. This line segment is

_____ cm long.

Draw a line segment $\frac{1}{2}$ as long.

Draw a line segment $\frac{1}{4}$ as long.

6. Solve.

Unit

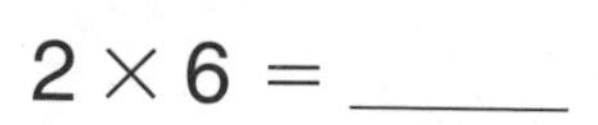

$2 \times 6 =$ _____

_____ $= 5 \times 8$

$16 = 2 \times$ _____

$5 \times 3 =$ _____

Date Time

Math Boxes 12.7

1.

Rule
$\times 3$

in	out
0	
1	
2	
3	
4	
	30

2. A brachiosaurus is 72 ft long, a diplodocus is 90 ft long, and a stegosaurus is 23 ft long. If they get in line behind one another, how long is the line?

The line is _____ feet long.

Number model:

____

3. Gary ran 800 meters. Kristen ran 628 meters. How much farther did Gary run?

He ran _____ meters farther.

Number model:

____

4. I bought a beach ball for \$1.49 and a sand toy for \$3.96. How much change will I get from a \$10 bill?

\$______

5. Write 3 ways to say 8:45.

____

____

____

6. Complete the Fact Triangle. Write the fact family.

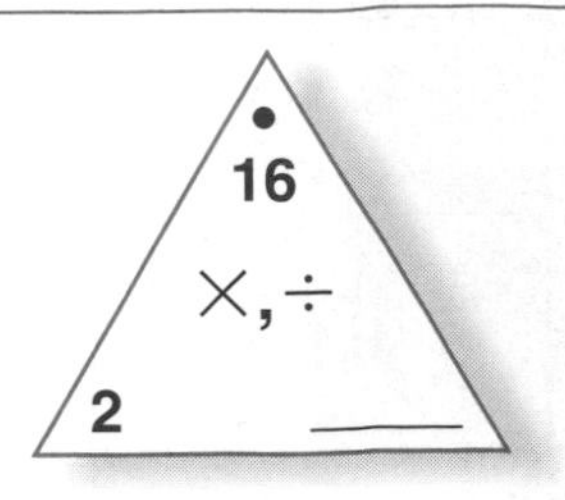

___ × ___ = ___

___ × ___ = ___

___ ÷ ___ = ___

___ ÷ ___ = ___

Date Time

Height Changes

The data in the table show the height of 30 children at ages 7 and 8. Your teacher will show you how to make a line plot for the data.

Student	Height	
	7 Years	8 Years
#1	120 cm	123 cm
#2	132 cm	141 cm
#3	112 cm	115 cm
#4	122 cm	126 cm
#5	118 cm	122 cm
#6	136 cm	144 cm
#7	123 cm	127 cm
#8	127 cm	133 cm
#9	115 cm	120 cm
#10	119 cm	125 cm
#11	122 cm	126 cm
#12	103 cm	107 cm
#13	129 cm	136 cm
#14	124 cm	129 cm
#15	109 cm	110 cm

Student	Height	
	7 Years	8 Years
#16	118 cm	122 cm
#17	120 cm	126 cm
#18	141 cm	148 cm
#19	122 cm	127 cm
#20	120 cm	126 cm
#21	120 cm	124 cm
#22	136 cm	142 cm
#23	115 cm	118 cm
#24	122 cm	130 cm
#25	124 cm	129 cm
#26	123 cm	127 cm
#27	131 cm	138 cm
#28	126 cm	132 cm
#29	121 cm	123 cm
#30	118 cm	123 cm

Date Time

Height Changes (cont.)

Use the line plot your class made to make a frequency table for the data.

Frequency Table

Change in Height	Number of Children
0 cm	
1 cm	
2 cm	
3 cm	
4 cm	
5 cm	
6 cm	
7 cm	
8 cm	
9 cm	
10 cm	

Date Time

Height Changes (cont.)

1. Make a bar graph of the data in the frequency table.

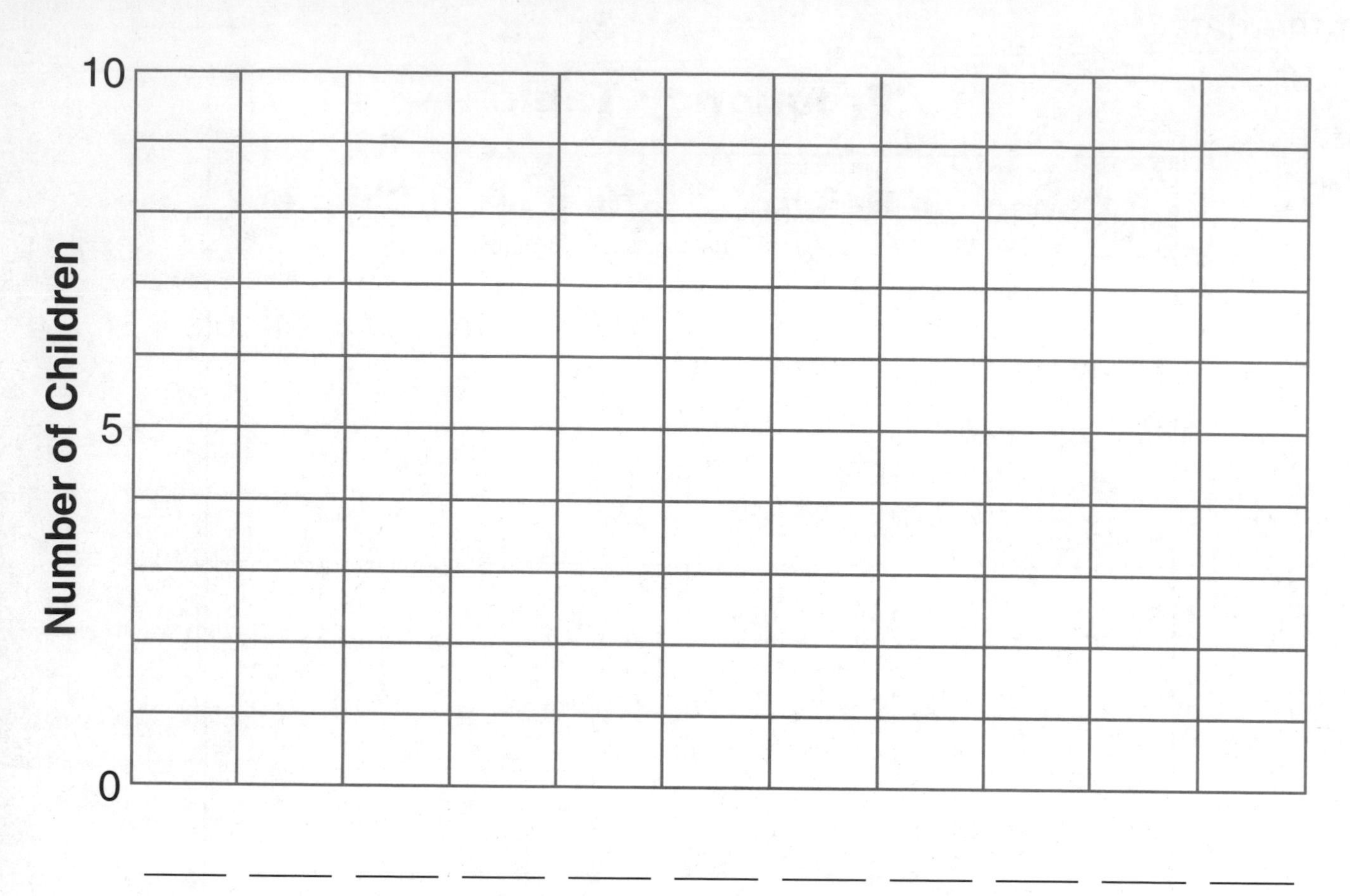

2. The smallest height change (minimum) is _____ centimeters.

3. The largest height change (maximum) is _____ centimeters.

4. The difference between the smallest and largest height changes (range) is _____ centimeters.

5. The middle value for the height change data (median) is _____ centimeters.

6. The height change that occurred most often (mode) is _____ centimeters.

Date Time

Math Boxes 12.8

1. The time is

____:____.

20 minutes later will be

____:____.

15 minutes earlier was

____:____.

2. 18, 15, 13, 17, 12, 17

Unit
peanuts

The median number of peanuts is ____.

The number of peanuts found most (mode) is ____.

3. Write a multiplication story by filling in the blanks. Solve the story.

7 ____________.

6 ____________ in each

____________.

How many ____________ in all? ____

4. Write the time in hours and minutes.

10 minutes past 12 ____:____

quarter-to 11 ____:____

half-past 7 ____:____

25 minutes to 8 ____:____

5. Solve.

Unit

$18 \div 6 =$ ____

(think $6 \times$ ____ $= 18$)

$25 \div 5 =$ ____

(think $5 \times$ ____ $= 25$)

$36 \div 9 =$ ____

(think $9 \times$ ____ $= 36$)

6. In 96,527, the value of

5 is ________.

6 is ________.

7 is ________.

2 is ________.

9 is ________.

Date Time

Table of Equivalencies

Weight	
kilogram:	1,000 g
pound:	16 oz
ton:	2,000 lb
1 ounce is about 30 g	

Symbol	Meaning
<	*is less than*
>	*is more than*
=	*is equal to*
=	*is the same as*

Length	
kilometer:	1,000 m
meter:	100 cm or 10 dm
decimeter:	10 cm
centimeter:	10 mm
foot:	12 in.
yard:	3 ft or 36 in.
mile:	5,280 ft or 1,760 yd
10 cm is about 4 in.	

Time	
year:	365 or 366 days
year:	about 52 weeks
year:	12 months
month:	28, 29, 30, or 31 days
week:	7 days
day:	24 hours
hour:	60 minutes
minute:	60 seconds

Money		
	1¢, or $0.01	
	5¢, or $0.05	
	10¢, or $0.10	
	25¢, or $0.25	
	100¢, or $1.00	$1

Capacity
1 pint = 2 cups
1 quart = 2 pints
1 gallon = 4 quarts
1 liter = 1,000 milliliters

Abbreviations	
kilometers	km
meters	m
centimeters	cm
miles	mi
feet	ft
yards	yd
inches	in.
tons	T
pounds	lb
ounces	oz
kilograms	kg
grams	g
decimeters	dm
millimeters	mm
pints	pt
quarts	qt
gallons	gal
liters	L
milliliters	mL

Date Time

Notes

Date

Time

Notes

Date Time

Fraction Cards

$\frac{2}{2}$	$\frac{1}{3}$	$\frac{3}{6}$	$\frac{1}{2}$
$\frac{3}{4}$	$\frac{2}{3}$	$\frac{1}{4}$	$\frac{2}{6}$
$\frac{4}{6}$	$\frac{0}{2}$	$\frac{6}{8}$	$\frac{2}{4}$
$\frac{4}{4}$	$\frac{2}{8}$	$\frac{0}{4}$	$\frac{4}{8}$

Date Time

×, ÷ Fact Triangles 1

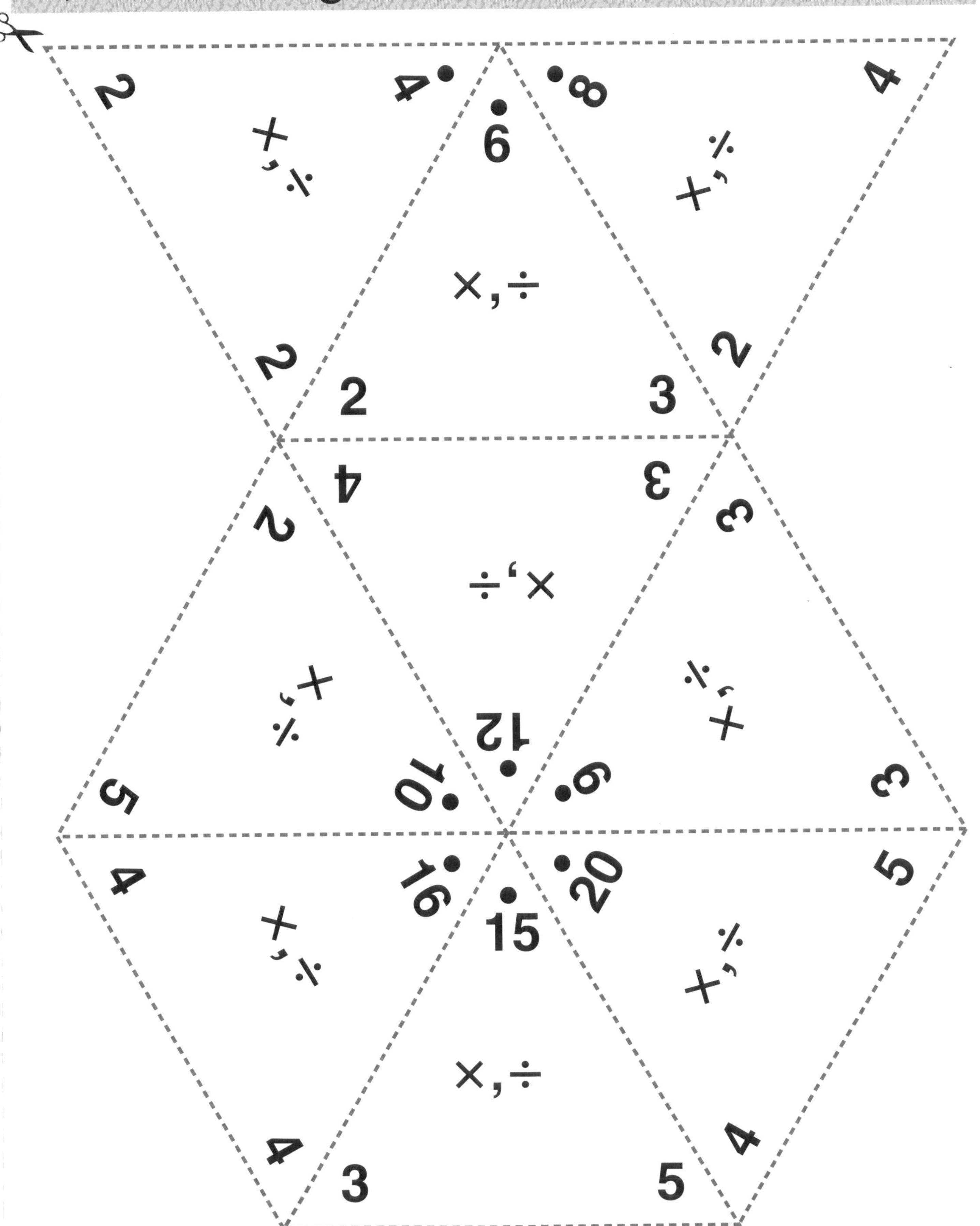

Date Time

×, ÷ Fact Triangles 2

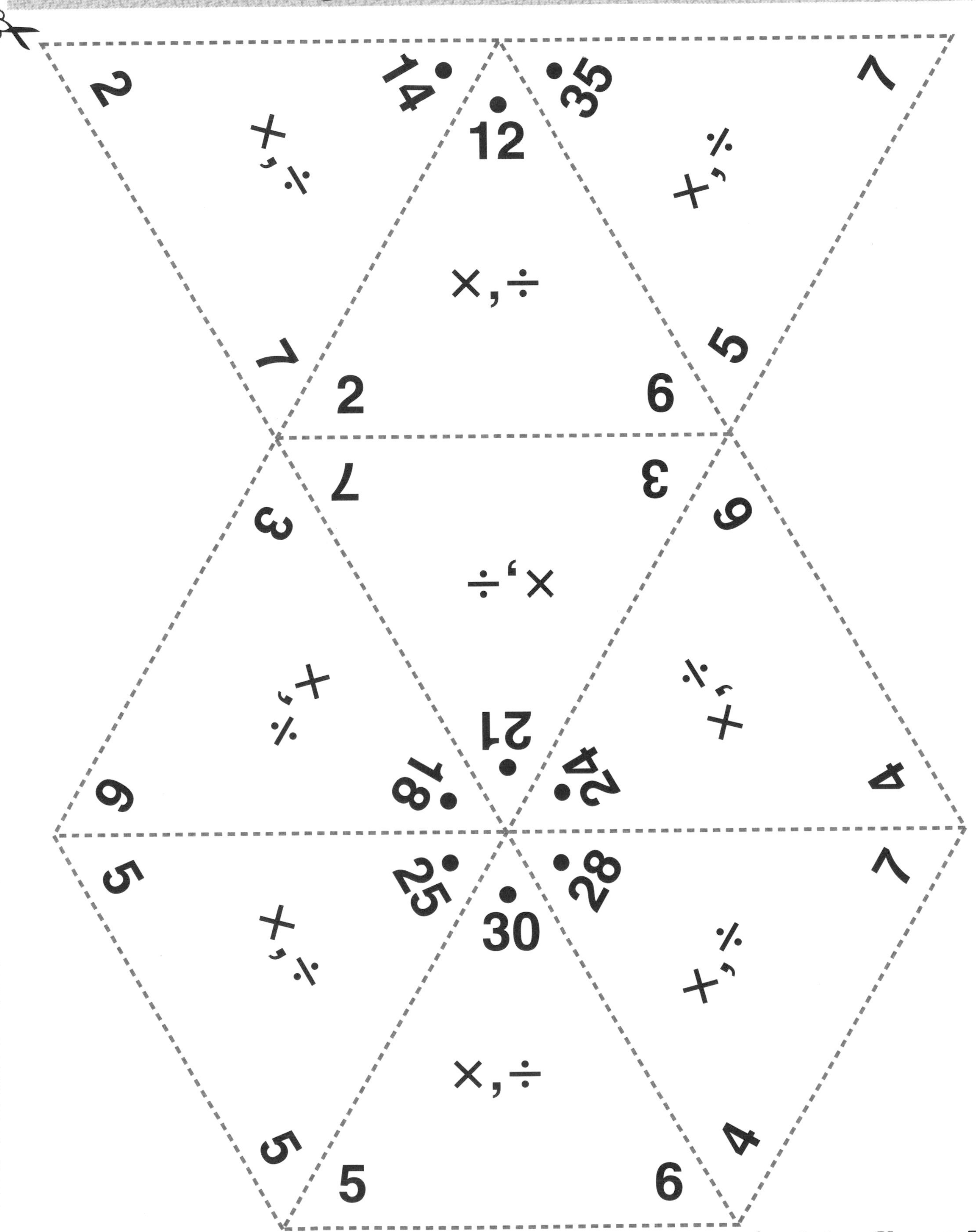

Date Time

×, ÷ Fact Triangles 3

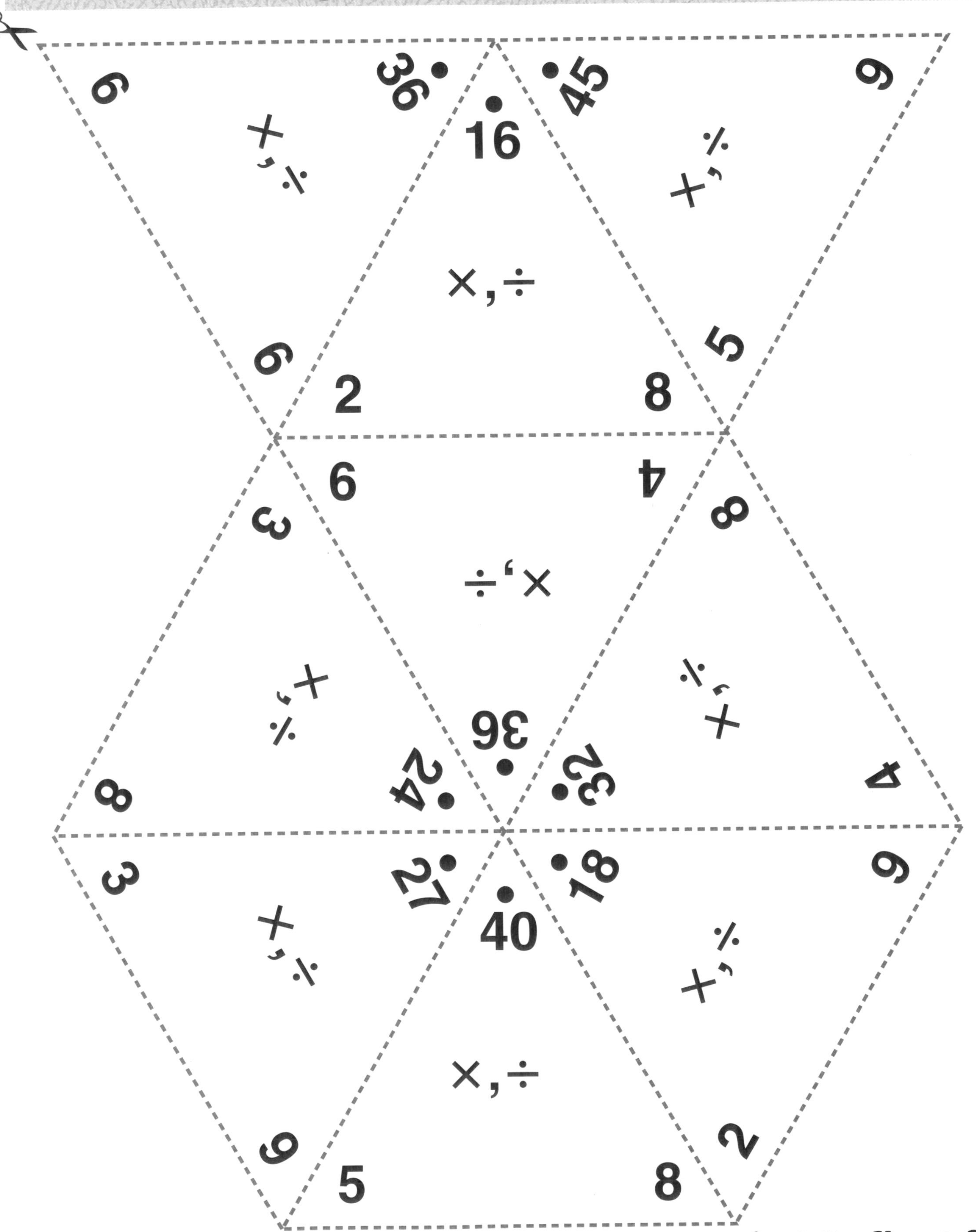

Date Time

×, ÷ Fact Triangles 4

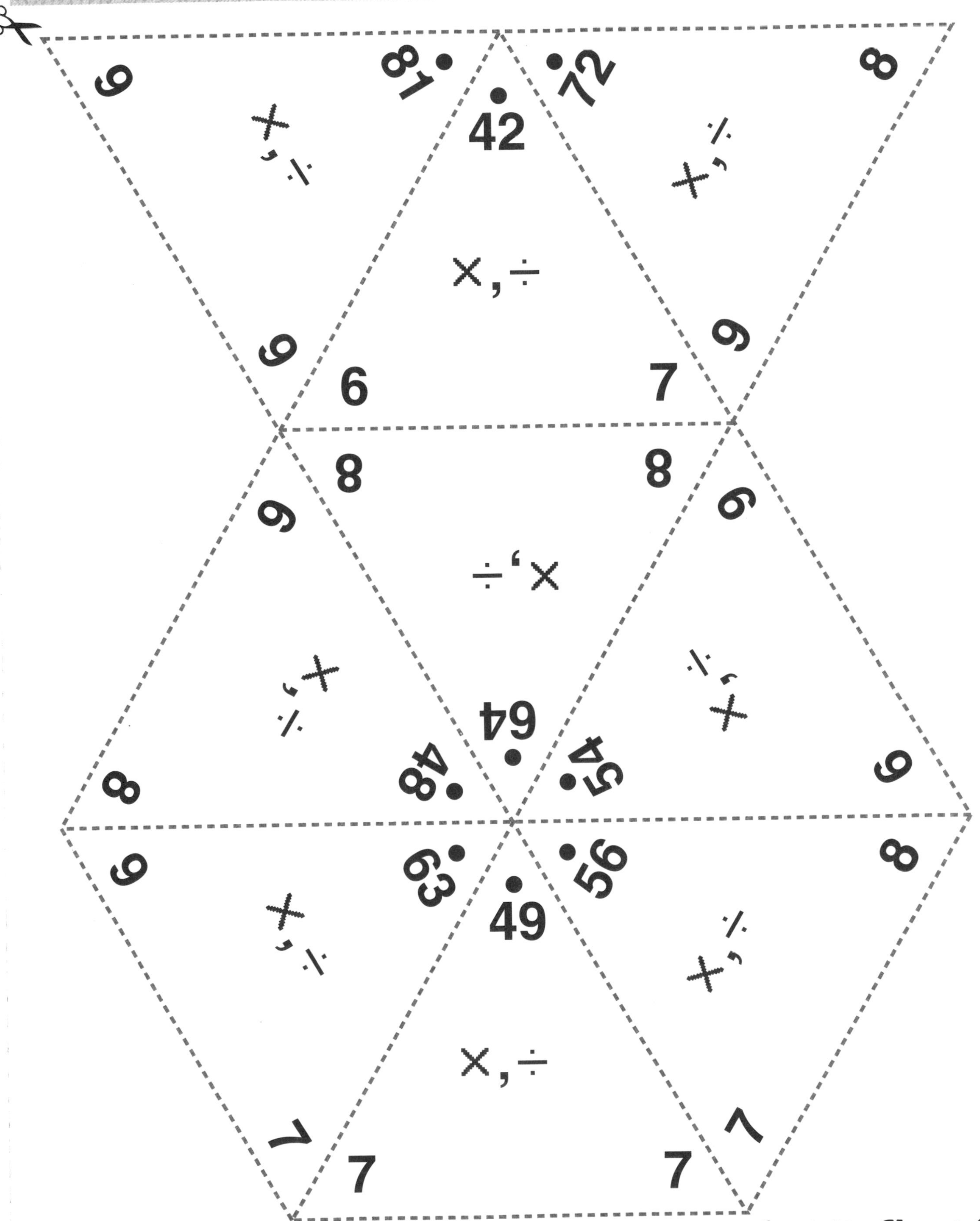

Use with Lesson 11.8.